Pennsylvania. State Geologist

A Summary Description of the Geology of Pennsylvania

Final Report Ordered by Legislature, 1891

Pennsylvania. State Geologist

A Summary Description of the Geology of Pennsylvania
Final Report Ordered by Legislature, 1891

ISBN/EAN: 9783744746625

Printed in Europe, USA, Canada, Australia, Japan

Cover: Foto ©berggeist007 / pixelio.de

More available books at **www.hansebooks.com**

GEOLOGICAL SURVEY OF PENNSYLVANIA.

FINAL REPORT ORDERED BY LEGISLATURE, 1891.

GENERAL INDEX

OF THE

FINAL SUMMARY REPORT

OF THE

GEOLOGY OF PENNSYLVANIA.

VOLUME I—Laurentian, Cambrian and Lower Silurian.
VOLUME II—Upper Silurian and Devonian.
VOLUME III—Parts I and II, Carboniferous and New Red.

J. P. LESLEY, State Geologist.

WITH AN APPENDIX CONTAINING

A LIST OF AND BRIEF GUIDE TO THE PUBLICATIONS OF THE SURVEY.

Compiled by

WILLIAM A. INGHAM,

Secretary of the Board of Commissioners.

HARRISBURG:
PUBLISHED BY THE BOARD OF COMMISSIONERS
FOR THE SECOND GEOLOGICAL SURVEY.
1895.

BOARD OF COMMISSIONERS.

His Excellency DANIEL H. HASTINGS, *Governor*,
and *ex-officio* President of the Board, Harrisburg.

WILLIAM A. INGHAM, ... Philadelphia.
HENRY MCCORMICK, ... Harrisburg.
CHARLES A. MINER, ... Wilkesbarre.
JOSEPH WILLCOX, ... Philadelphia.
LOUIS W. HALL, ... Harrisburg.
SAMUEL Q. BROWN, ... Pleasantville.
CHARLES H. NOYES, ... Warren.
W. W. H. DAVIS, ... Doylestown.
ECKLEY B. COXE, ... Drifton.

SECRETARY OF THE BOARD.

WILLIAM A. INGHAM, ... Philadelphia.

STATE GEOLOGIST.

J. P. LESLEY, ... Philadelphia.

GENERAL INDEX.

A.

Abbott coal bed, Wilkes-Barre, ... 1997
Academia anticlinal, ... 989
Ackley ore bank, ... 378
Adair's quarry, limestone, ... 981
Adams county, ... 142-144,455
    "   "   mines, limonite, ... 225
    "   "   quarries in South Mts., ... 556
    "   "   South Mts., ...29,62,70,163
    "   "   map in C2, ... 242
    "   "   map in D5, ... 321
    "   "   trap dykes, ... 455
    "   well, ...1486
Adam Stahl tract, diamond drill hole, ...2100
Adelaide mine, S. W. Pa. Pittsburgh coal, ...2491
Adirondack Mts., ...61-2,113,158,499,535,671
    "   region, ... 44
    "   mines, ... 115
    "   rocks C. E. Hall's report on, ... 109
Adrian mines, Jefferson co., ...2297,2302
Aethiopia old drainage, ... 11
African worm burrow casts, ...1347
Agassiz, Louis, ...51,774,1627
Agey mine, Blairsville, ...2331,2334
Ahl mine, Franklin co., ... 229
   " prospecting pits, ...1231
Aims' slate quarry, ... 590
Alabama, exposures in, ... 40
    "   Great Valley to New York, ... 553
Alaska colliery, Skidmore and Holmes beds, ...2065-6
    " coast, banks of the, ... 36
Albany county, New York, Knowersville gas well, ... 560
    " mine, Lisbon basin, ...2521
Albright colliery, Tremont basin, ...2118-24-25

Alden colliery, Nanticoke, ...2003-4-5-6
Alexander limestone quarry, ...371,1121,1002
Allegany county, New York, ...1372,1422,1544
Allegheny county, ...1542-43
" " No. XIII in, ...2406
" " " XIV in, ...2442
" " " XV in, ...2524
" or Cumberland Mountain, ...1793
" Mountain, ...1651,1725,1785,1809,1858,2220,2244
" " coal field, ...858,1857
" " escarpment, ...849,1354,1573,1837-9,1840-65
" " coal prospectus, ... 42
" " counties, ...1860
" " gap at Lock Haven, ...1723
" " No. IX Catskill along the, ...1616
" " No. XI Mauch Chunk along the, ...1840
" " plateau, ...1621,1842,1861
" " " of Lycoming county, ...1423
" " region, ...2287
" " summit tunnel, P. R. R., ... 433
" river coal series, ...2316 Chap. CXXIV
" river, ...681,1190-1,1372,1451,1462,1482-89-91,1502-5-7-31,1544
" oil region, ...1747
Allegheny-Washington-Butler region, ...1544
Allegrippus ridge, ...1257,1554-5
Allen cement, ...306,337-40
" Robert, well section, ...1897
" quarry, ... 990
" " section, ...1514
Allentown levels, ...272-3
" furnaces, ...301,1065
" Iron Company's mine, ... 233
Allequippa mine section, ...2548
Allison mine section, ...2541
Allsworth pit section, ...2500
Allwein's quarry, limestone, ... 321
Alms' coal mine, Indiana county, ...2463
Alpha coal bed, ...2024-27-29-36-37-39-42-44-47
Alpine mountain range, ... 82
" regions, ... 432
Alps, ...58,59,76
" Tyrolean dolomite, Austrian survey, ... 327
" Jurassic, Cretaceous and Tertiary slopes, ... 52
" overthrust faults, ... 703
Altamont collieries, ...2080
Altfather mine, ...2244

INDEX FINAL SUMMARY REPORT.

Alton basin, ...1876-79-80
"   coal bed, ...1877-82-83-84-85
"   coals of No. XII, ...1881,2282
"   "   group, ...1873-76,1878,2280
Altoona, Sander's section at, ...1681
American mine, ...2547
"   series, ...1312
Amity mine, ...2548
Analyses, anthracite, ...1927
"   "   Schuylkill Dauphin-basin, ...2146
"   "   bed B, Bernice basin, ...2012
"   "   bony coal, ...1928
Anchor cement, ...337-9
Anderson's quarry, ...1508
Anita (West Eureka) mines, ...2302
Ankeny mine, ...2254
Annora colliery, ...1979-80
Annville limestone quarries, ...314 to 320
"   level near, ...274
Ansonville bank section, bed C', ...2215
Anthony mine, ...2355
Anthracite formation, No. XIII, Chaps. CXVIII to CXXIII,....1916-2152
"   composition of, ...1926
"   survey, ...1594
"   "   plan of. 1925
"   "   scope of final report, ...1939
"   region, ...1857-61,1917-24,2154-63
"   "   area of, ...1918
coal fields, contents of, Chap. CXXIII, ...2147-51
"   "   "   coal remaining in, ...2152
"   "   "   production, ...1916,2152
"   beds originally bituminous, ...1919
"   coal measures, ...1917,1923
"   "   "   thickness. 1924
"   "   "   structure, ...1924
Anthracite, commercial importance, 1916; specific gravity, 1928; vegetable origin, 1929; mining methods, 1932; mine sheets, 1935; cross section sheets, 1936; columnar sheets, 1936; topographical sheets, 1936; miscellaneous sheets, 1937; reports, 1937;Northern coal field, chap. CXIX, 1940; refuse in coal banks, 1952; structure, 1947; coal measures, 1951; sub-divisions, 1952 to 2013; limestone beds 2020; peat bog mineral, 2021; Eastern Middle coal field, chap. CXX, 2022; mine sheets, 1942; coal measures, 2024; sub-divisons, 2025 to 2046; Western Middle coal field, chap. CXXI, 2049, 1943; mine sheets, 1943; maps, 1945; sub-divisions, 2052 to 2066; Southern field, chap. CXXII, 2072; maps, 1944; structure, 2074; coal measures, 2075; condition of beds, 2076; sub-divisions, ...2077 to 2138

GEOLOGICAL SURVEY OF PENNSYLVANIA.

Anticosti, ...617,717,877-9
Antrim mines, ...2273
Apollo axis, ...2352
Appalachian region, 17, 24, 37, 89, 197, 380, 627, 804, 1316, 1430; area, 1423,
    2153; outcrops, 1370; faults, 352; basin 1312, 1423, 1431; land, 1426;
    sea, chap. V, 35 to 52, 437, 525-9, 539-40, 627-73, 721-36, 901-10, 1015,
    1141-87, 1451; sea bottom, 947; gulf, 1426; shore, 1426; geologists,... 625
Apple mine, ... 234
"   William; opening, ...2253
Appleman's quarry, ...949-53,1025
Appleton's limestone quarry, ...1074
Ararat mountain, ...1455
Archaean rocks, chaps. VIII, IX, 63, 74; are sedimentary? chap. X,..... 95
Archbald coal bed, ...1963
"   colliery, ...1970-72
"   axis, ...1960
"   pot-holes, ...2016
Arcola coal, ...2627
Argyle opening, ...2227
Arizona, ... 194
"   Valley of Death in, ... 52
Arkansas river, grand canon, ... 756
Armenia mountain, ...2264
Armstrong county, No. XII, 1890; No. XIII, 2343; No. XIV, 2427; No.
    XV, 2469; coal analyses, 2359; fire clay analyses, 2361; limestone
    analyses, ...2360
Armstrong valley anticlinal, ... 963
Armstrong's oil wells, ...1507
Arnot mine section, ...2273
Ash, John S. finds fossils in Norristown shales, ...2612
Ashburner, Charles A. 489,560,1486-96,1541,1926-35-38-83,2008-16-17-21
Achburner, Charles A. section Sideling Hill, 1330; section Aughwick
    valley, 995-6; section, McKean county, 1417; section, Mifflin
    county, 989; section, Catskill, 490; section, general, 813, 1613;
    section, East Broad Top, 1330; section No. IV, 631; section, Lo-
    gan's gap, 645; section, Rockhill gap, 645; well records, 1191, 1487;
    classification, etc., ...1481 to 1485
Ashcroft mine, ...2236
Ashland, depth of Mahanoy basin at, ...2050
"   gap, elevation of, ...2049
"   furnace, ... 217
"   Company's bank, ...184,217
Ashley, ...1983
"   colliery and section, ...1821,1989,1992
Askam diamond drill bore hole, ...1987-98
Aspdin process, ... 339

Asylum quarry, ... 1193,1530
Atlantic coast, rivers and lowlands, ... 18
"   ocean, area of, ... 1032
"   "   southern, water level of, ... 723
"   "   dredgings of the Albatross, ... 1015
"   colliery, ... 2535
Atlas mine section, ... 2492
Auble coal bank, ... 1998
Aughwick valley section, ... 995-96
"   synclinal, McVeytown outcrop, ... 1098
Auvergne basaltic beds and columns, ... 14,98
Avery, Smith &, bank, ... 239-51
Avoca, ... 1974
Avondale colliery, ... 1990
Axial directions, cause of diversity, New Red area, ... 2601

B.

Babbett's mine, ... 2539
Babylon colliery, ... 1977
Bache mine, ... 2270-3
"   "   bed B, ... 2189
Back coal bed, ... 2034
Bade's mine, ... 234
Bahama banks. 1317
Bahrrer's quarry, ... 481
Bailey's mine, ... 2390
"   quarry, ... 1004,1514
Baird's quarry, ... 1003
Baker mines, ... 846,853-4,1002,1001,1119,2230
"   and Trout's mine, ... 1000
"   quarry, ... 1001,1119,1801
Bakerstown coal, ... 2408
Bald mountain, ... 269
Bald Eagle mountain, 645-57-59-81-2-4-5-95-6 859-60; mines along foot, 372; faults, 860; structure and sections, 860 to 865; outcrop, 659-67; gaps, 631, 1572; gaps, measurements, 645-7-9; anticlinal, 483, 705-37, 937-44, 1124-78, 1262; ravines, 1003; range, ... 669
Bald Eagle valley, ... 859,1235,1355,1617
"   "   "   No. VI in, chap. LXXII, ... 999
"   "   "   Marcellus ore, ... 1234
"   "   furnace, ... 858
"   "   "   fault, ... 860
Baldwin's mill opening, ... 2575
Baldy's quarries, ... 947-55
Ball, T. opening, ... 2299
Balliet mine, ... 234-5,341-3

Ballow bank,	369
Baltimore coal bed,	1991,2004
" colliery,	1987 to 91
" quarries on Jones' and Gwynn's falls,	130
Bamford zinc mines, Lancaster county,	440-3
Bangor slate belt, history of,	582
Bangor quarries,	546-84-90-91-92
Banner mine No. 1,	2552
Barber, J. & Co.'s, mine,	233
Barber & Almy's mine,	234
" Alney's mine,	346
" quarry,	319
Barlow & Day mine,	373
Barclaugh coal bed,	2112
Barclay basin,	1839,2185,2263-4
" " bed A,	2268
" mines,	1833
" " sections,	2267
" " analyses,	2269
" Rutherford &, slope,	1176
Barge, Beach & Co.'s quarry,	603
Bargeddie mine,	2521
Barium in formation No. II,	Chap. XXXVII, 436
Barnhard, N., opening,	2559
Barnhart, H., opening,	2510
Barhart opening,	2488
Barnet coal bed,	2173
Barnett's rocks,	1215
Barnum colliery	1978
Barree Forge,	934
" limestone group, 825; lower shales,	826
" forge mining district,	828
" " section above,	837
" upper shales,	825
Barrens' group in lower Silurian limestone,	378
" The,	104,369-70-4,429
" " of Centre county,	498
Barren Measures, No. XIV,	2247-61-63-92
" " No. XIV or Pittsburgh series,	Chap. CXXV, 2406
" ridge,	1091
Bartholomew's quarry,	1773
Barto mine,	269
Barytes in No. VI,	915
Bass shaft,	1177
Bassler quarry,	314
Bastard and Bossardville limestomes,	948

Bastian mine,	234
Batdorf & Beaver quarry,	318
Bath,	599,1387-99
" quarry, near,	308
" road,	596
Bauman's bank,	217-18
Bayard's shaftings,	2142
Beach, Barge & Co.'s quarry,	603
Beaghley bank,	2261
Beall's mill,	1222
" " anticlinal,	1091
Beam bank,	394
" " section bed B,	2215
" mine,	2260
Bear creek seam,	2274
" " well,	1487,1729-35,1840
" ridge colliery,	2060
" " overturned axis,	2053
" " slope,	2116
" run colliery,	2057
" valley colliery,	2070
" " shaft,	2136
" " slope,	2137
Beartown mine,	229
Bear's George bank,	227
Bear's Head, elevation of,	2049
Beaty mine, bed E,	2331
Beaver county,	2439
" " No. XII in,	1912,2396
" " No. XIV in,	2440
" Lawrence and Mercer Co.'s, No. XII in,	1914
" and Shenango valleys, No. XII in,	1914
" bank,	359
" quarry,	317
" dams,	398,1000
Beaver's S. bank,	211
Beaver Brook collieries,	2043
" falls well,	1769
" Meadow basin,	2041
" " colliery,	2043
Beaver-Rutherford-Paxtang limestone belt,	224
Bechtal's bank,	134,216-7
Beck banks,	233,378-91-98
Beck's Run mine,	2546
Beech Tree mines,	2298,2306
Beechwood colliery,	2105-8-10

Beckley's quarries, ... 315
Bed A, Northern coal field, 2001-10; B, Panther creek basin, 2088; Bernice basin, 2009-13; C, Panther creek basin, 2088; D, Panther creek basin, 2090; D, analysis of coal, 2210; D, analysis of ash, 2211; D, analysis for phosphoric acid, 2211; E, in Panther creek basin, 2089; F, or lower red ash, 2091; G, or upper red ash, ...2092
Bedford county, 819 to 823, 837 to 849, 1000, 1113 to 1118; measurement of No. IV in, 631, 661; fossils of V, 895; and Fulton No. V., Chap. LX, 838; No. VI, Chap. LXXI, 983 996 to 998; No. VIII, 1103, 1114 to 1118; No. VIIIa, 1189; No. VIIIc. 1281; Portage, 1356; Chemung-Catskill fossils, 1562; Catskill, 1613; red shale, 1775-79; No. XII, ...1864
Bedford synclinal, ...1281
" gaps near, ...631,661
" Marcellus ore, ...1233
" county limestone coves, ...,................... 425
" springs, ...997,1115,1117
" pit (old), ...1229
Beech Creek, or First basin, ...2199
Beelor trap dyke 215
Belleview colliery, ...1971
Bellevue mine, ...2533
Belgium, ...1589
Bellefonte dam, ...431-2
" anticlinal, ... 433
" cross section, ... 370
" outcrop, ... 619
" gap, ...599,647,659,858
" glass works, ...1122
Bell's tunnel, ...2097-8,2100-1
Belmore colliery, ...2064 to 66
Bellwood mine, ...2546
Benezette dry hole, ...1731
Ben Franklin colliery, ...2060
Bennett coal bed, ...1991,2006
" colliery bore hole, ...1987
" bank, ...2325
Bennington section, ...2221
" mines, ...2224
Bentleysville mine, ...2536
Benton mines, ...2241
Berks county, ...267-8,289-90-1-2,543
" " trap dyke, ...451-4,464,564
Berlin-Salisbury basin, ...2244-5
" coal bed, ...2420
" limestone group, section, ...2421

Bernice coal basin, ...1837,2008,2189
" colliery, ...2007
Berry mountain Pocono in, ...1655
Berwind-White Coal Co.'s mines, ...2297
Berwind shaft, ...2298,2304
" " bed D section, ...2214
Best mine, ...2302
Bethel Coal Company's shaft, ...1907
" mine, ...2250
Biesecker, Noah opening, ...2252
Big Black creek basin, ...2032
" Bottom bed analysis, ...1852
" coal bed, ...1971
" flat coal opening, ...2145
" Lick mountain axis, ...2124
" Mine colliery, ...2063
" red (Bedford) formation, ...1775
" Soldier mine, ...2297,2305-6
" " run mine drift, ...2303
" Trough creek coal basin, ...2181
Billin, C. E. measurement of IV in Centre county, ... 631
" " sections in Mifflin county, ... 668
Billy Best coal bed, ...2083
Birmingham Coal Company's mine, ...2517
Bituminous coal (see Carboniferous Formations), XIII, XIV, XV,
  XVI, ...2153
Black creek basins, 2034-5-6; Diamond colliery, 1988-92-4, 2065; Diamond mine, 2550-2-3; Hawk mine, 2558; Heath coal bed, 2122-31; Heath tunnel, 2120; Lick coal field, 2221-38; Lick seam, 2220-22-39; Log mountain anticlinal, 1229-81; Log mountain gap, 669, 817; Log mountain gash fault, 713; Log mountain mines, 818; Log valley, 303, 419-22, 523, 701; Log valley limestone and ores, Chap. XXXV, 419; Marble in No. IIc, 482, chap. XL, 482; Marble in No. VI, 484; mine coal bed, 2115-23; Ridge colliery, 2035; Springs gap, section, ...2143
Black Walnut quarry, ...1593
Black's bank, ...2563
Blackburn mine, ...2549
Blackwood colliery, ...2118-20-22-23
Blair, A. A. analysis, ...1034
Blair county, 647, 849-50-3, 1116 to 1120, 1234, 1332-55, 1552, 1841; Nittany valley, 17; Canoe mountain, 661; Morrison's cove 395, 521; gaps, 551; section, 851, 1616-19; Sinking Valley zinc and lead mines, 444; Oxide works, Birmingham, 442; Limonite mines, 341, 568; measurements of No. IV, 631-45; No. V, Chap. LXI, p. 849; No. V fossils, 893; No. VI,Chap. LXXII, p. 999 to 1002; No. VII 1103-18; Hamilton fossils, 1366; Portage, ...1355

Blair open pit, ...1002
Blairsville, 1511-73; gap, 28, 1715; wells, 1543; basin, 2159, 2316,-28-69, 2465; DuBois basin, 2298,2302; Connellsville basins, 2472, 2565; and Saltsburg basin No. XIV, 2426; and Saltsburg axes, 2432; axes,..2473
Blank mine, ...233-4
Bleakley quarry, ...1517
Blessinger's bank, ... 212
Bloomfield mines, ...395,404-14-16-17,1119
" anticlinal, ... 409
Bloomsburg gap, ... 737
" red shale, ... 805
" furnaces, ... 951
" section, ...747,1260
" mine, analysis, ... 749
" Iron Company's R. R., ...1260
Blossburg, ...1418-56-8-66-7
" anticlinal, ...1443
" synclinal, ...1457
" basin, ...1448,1871,2189,2263-4-9-76
" " Monkey ledge in, ...1803
" mines, ...2272
" mountain, ...1721
Bloss, G. farm, section and analysis, ...2299
" vein, analyses, ...2265-72
Blue Ball fire clay, ...1868
" Jay well, ...1491
" mine, ...1877
" mountain quarries, ...606-7
" Ridge, ...70,78,144-6,291,1032,1375,1962
" vein quarry, ...579-80,605
Boalsburg-Fillmore cross-section. 371
Bolivar fire clay analysis, ...2366
Bollenfields school, opening, ...2561
Bone cave in No. VI, ... 916
Bony coal analysis, ...1923
Borland, W. opening, ...2301
Boston colliery, ...1976-9-93-4
Bostonia mines, ...2351,2429
Bottom coal bed, ...1959
Bowden, J. H. 2017
Bowkley coal bed, ...1996
Bowman mine, ...2253
" Hill trap, ...2619-22
Boyd mine, ...2523
" Hill gas well, ...1713-99
Boyer opening, ...2488

Boyertown basin, ...2599
Bradford county, ...1418 to 1428,1434 to 1439,1832
"   "   fossils, ...1468
"   "   fish beds, ...1456-67
"   "   Chemung, ...1443 to 1452
"   "   No. XII, ...1871
"   "   No. XIII, ...2263
"   oil fields, ...1191,1372,1481-4-7,1862
"   " rocks, ...1733
"   " sands, ...1484.1747
Brady's Bend anticlinal, ...1876,2373-84
"   "   axis, ...2403
"   mine, ...2341
Branchdale colliery, ...2124
Brandt mine, ...2254
Brandywine creek, ...91,201-2
"   "   gorge, ...120,131-2
"   "   gaps, ...177-8
Brazilian diamonds, ... 58
Breniser mine, ...2472
Brewery coal bed, ...2116
Bridge colliery, ...1870
Bridgeport sandstone, ... 761
Bridgeville mine, ...2541
Brindle & Harbaugh mine, ...2487
Brisben colliery and coal bed, ...1973
Brisbin, William, bank, ...2513
Brister, H. opening, ...2559
Broadford mines, ...2476
Broadhead's creek gap, ...923,1055
"   "   No. VII on, ...1054
Broad Mountain, ...1813
"   "   basins, ...1811,2077
"   "   Coal Company's explorations, ...2080
Broad Top mountain, 681, 706, 1574; basin, 295, 681, 820, 1574, 1601-9, 1861, 2154-63-5-75, 2409; general section, 2160; anticlinal 2167-80; synclinal, 1107, 1610; East section, 1330; Catskill around, 1609; coal field, No. XI, in, 1833, Chap. CXXV; coal field No. XII in, 1862; general section, 2168; East coal field, ...2180
Brookside colliery, ...2127-8-9
Brockwayville district, ...2298
Brookville anticlinal, ...2373
"   coal, ...2383,2402
"   " A, ...2222-53-94,2313-17-57-78
"   " A in Snow Shoe basin, ...2199
Browser, A. opening, ...2559

14 GEOLOGICAL SURVEY OF PENNSYLVANIA.

Brown, Dr. A. P. lithology of the trap. 2623-37
" P., opening, ...2332
" basin, ...2172
" mine, ...2303
Brown's colliery, ...2112
" drift, ...2232
Brownstone quarries, ...2627
Brubaker region, ...2221
Brushy mountain, ...1631
Buchanan mine, ...2324
Buck mine, ...2487
Buck mountain, 1854; colliery, 2033-9; coal bed, 2002; coal bed in Eastern Middle coal field, 2026 to 2029-33 to 39, 2043 to 47; coal bed in Western Middle field, 2055-60-5-9; coal bed in Southern field, 2096, 2108-20-30-36; coal bed in Broad Mountain, 2079-80; coal bed in Heckscherville basin, ...2083
Buckingham opening, ...2559
Bucktail mine, ...1884,2259
Buckville colliery, ...2096-8,2102
Buffalo valley, ...871
" " synclinal, ...985,1271-4
" " anticlinal, ...873
" mountain, ...1657
" mine, ...2550-2
Buhrstone iron ore, ...2162,2294,2344-92
Building stone, ...2627
Bulger axis, ...2393
Burgen mine, ...2556
Buried valley, Newport creek, ...2017
" " Wyoming, ...2017-18
Burlington limestone, No. VIII, ...1447,1791
Burnside colliery, ...2070
Butler county, No. XII, ...1896-98
" " No. XIII, ...2380-83
" " No. XIV, ...2436
" oil sand, ...1711
" Coal Company, ...3673
" colliery, ...2109-11
Buttonwood anticlinal, ...1986
" shaft, ...1997
Buttsville mine, ...1879
Butzbach Landing bore hole, ...1196

C.

Cale Hollow, 398; bank, 387; mouth of, 398; anticlinal ridge, 382; range in Huntingdon county, ... 394
Caledonia basin, Cameron and Elk counties, ... 1883,2280,2282
" 1163, 1240; furnace, 148, 150; lands, ... 149
" mine, Pittsburgh bed at, ... 2554
Caldwell's opening, Pittsburgh bed, near Smithfield, ... 2496
" section of bed C' at, ... 2215
Calumet mine, Pittsburgh bed at, Connellsville basin, ... 2488
Cambria county, No. XI in, ... 1841
" " No. XIV in, ... 2412
" " No. XII m, ... 1865,1869
" " No. XIII in, ... 2249
" mine; ore in No. V, 845; coal mine in Bedford county, ... 2171
" " Kelly coal bed section in, ... 2170
" Iron Company's Cresswell quarry, ... 1001
" " Company, 407, 836, 852, 855, 857; mine, 837; mine analyses, ... 413
". " Company's slope near Frankstown, ... 893
Cambrian fossil life, Chapter XVIII, ... 192
Camden mine, Pittsburgh coal at, ... 2548
Cameron and Elk counties, district, ... 1839
Cameron-Caledonia basin, ... 2281
Cameron coal basin, ... 2280
" Coal Company, property of, ... 1881
" coal basin, of Cameron county, ... 1471
" colliery, Shamokin gap, ... 2068,2070,2071
" county, No. XII in, ... 1866,1867,1881
" county, No. XIII in, ... 2279
" county (oil sands), ... 1486
" well and section, ... 1488
Campbell, farms, openings in Pittsburgh bed, ... 2488
" limestone mined by, 1447; quarry, 523; well, ... 490
Campbell's A. mine, bed B in Ligonier basin, ... 2327
" ledge, black shale bed, ... 1950,1959,1967,1975,2001
" " and Coxton section, ... 1607
" mine, Philson and Coleman bed, No. XIV, ... 2425
Canada, Hamilton rocks in, ... 877
" mountains of (Laurentian belt), ... 44,61,62,108
Canadian Azoic rocks, ... 77
" New Red and Palaeozoic, compared with ours, ... 2614
Canoe Mountain gap measurements, ... 645
" " synclinal, ... 488
" Valley anticlinal, ... 403,408,409
" " limestone and ore, Chapter XXXIV, ... 401

Canty's quarry, Third Sand, Erie county, ...1517
Cape Cod, glacial area from, to Manitoba, ... 51
Capon axis, ... 756
Capouse colliery big bed, Northern coal field, ...1971
Carbon Coal Company's mines; Mercer lower limestone, Mercer county, ...1904
" county, Catskill in, ...1585
" " No. VI in, Chapter LXVII, ... 936
" " section, Catskill rocks, ...1584
" Metallic Paint Company, ...1176
Carbondale, description of coal measures about, ...1952
" Forest City division, ...1952
" Scranton region, ...1576
Carboniferous age of peat bogs, ...1631
" fire clay beds, ...2202
" series; Pottsville conglomerate base of, ...1857
" system, ...1829,2153,2345
" No. XI in Somerset county, ...2248
" No. XIII, Lower Productive or Allegheny river coal series, Kittanning group in Marion-Saltsburg basin, ...2343
" " " beds C', C and B, ...2352
" " " Cameron, Elk and Forest, ...2279
" " " Jefferson, ...2292
" " " Indiana, ...2315
" " " Armstrong, ...2343
" " " Westmoreland and Fayette, ...2362
" " " Beaver county, ...2396
" " " Allegheny county, ...2403
" " " Clarion county, ...2373
" " " Butler county, ...2380
" " " Mercer county, ...2387
" " " Lawrence county, ...2391
" " " Cambria county, ...2219
" " " Somerset county, ...2241
" " " Bradford and Tioga counties, ...2263
" " " Potter county, ...2275
" " " McKean county, ...2277
" " " Sullivan and Lycoming counties, ...2134
" " " Clinton county, ...2190
" " " Centre county, ...2194
" " " Clearfield county, ...2205
" " " Lower Productive Measures, Chaps. CXXIV, and XXXV, ...2155,2165
" No. XIV, Barren measures or Pittsburgh series in Ligonier valley, ...2424

Carboniferous No. XIV, Armstrong county, ...2427
"   "   "   Westmoreland and Fayette counties west of
                     Chestnut Ridge, ...2432
"   "   "   Butler county, ...2436
"   "   "   Cambria county, ...2412
"   "   "   Somerset county, ...2414
"   "   "   Johnstown Trough, ...2421
"   "   "   Jefferson county, ...2422
"   "   "   Indiana county, ...2423
" •   "   "   Lawrence and Beaver counties, ...2439
"   "   "   Allegheny county, ...2442
"   "   "   Greene and Washington counties, ...2446
"   No. XV, Upper Productive or Monongahela river series,
                     in Lisbon gas coal basin, ...2501,2515
"   "   "   Allegheny, Washington and Greene counties,..2524
"   "   "   Pittsburgh district, ...2538
"   "   "   Armstrong county, ...2469
"   "   "   Ligonier valley, ...2470
"   "   "   Westmoreland and Blair counties, Connells-
                     ville coking basin, ...2472
"   "   "   Westmoreland county, Greensburg basin, ...2497
"   "   "   mines along Saw mill run, ...2545
"   "   "   "   on Pool No. 1, ...2546
"   "   "   "   "   "   No. 2, ...2543
"   "   "   "   "   "   No. 3, ...2550
"   "   "   "   "   "   No. 4, ...2554
"   "   "   "   "   "   No. 5, ...2557
"   "   "   "   "   "   No. 6, ...2560
"   "   "   "   "   "   Nos. 7 and 8, ...2562
"   "   "   Salisbury basin, Somerset county, Chapter
                     CXXVII, ...2461
"   "   "   Character of coal, ...2464
"   "   "   Indiana county, ...2465
"   No. XVI, Upper Barren Measures, Chap CXXVIII, ...2525
"   "   "   in Washington county group, ...2569
"   "   "   Greene county group, ...2580
Carll's colored geological map of Warren county, ...1491
"   Elk oil sand group, lower Chemung rocks, ...1485
"   reports (on the oil fields), ...1501
"   section at Great Bend, Warren county, ...1416
"   "   at Warren, ...1492
Carmalt opening; coal bed D, Young township, Jefferson county, ...2301
Carrick Furnace Company; Path valley limonite iron ores, ...359,360
Carroll quarries, Venango oil group. ....1512 to 1517
Carolltown-Patton field; Cambria county coal field, ...2221

2

Carter's, O. S. C. notes of coal in New Red, ...2620
"   " notes of deep borings for water, New Red, ...2634
Cascade Cr. 1440, 1556; fall, 1439; Starruca creek and Cascade section, 1440
    mines; Kittanning coal fields, Elk county, ...2284
Cashup well, 1537; record, ...1539
Castle mountain (limestone), ... 192
Castle Shannon mines, Pittsburgh coal at, ...2545
Cassel's quarry, Dauphin county, ... 321
Catawissa section, ...1604,1552
" -Bloomsburg section. 1260
" and Nescopec mountains, ...1651
" -Rupert section, ...1260
" -Sunbury basin, ...1262
" synclinal, ...1258
Catskill No. IX, Chapters CVI to CX, 1567 to 1628; in the northeast
    counties, Chapters CVII, CVIII, 1576, 1585; red
    color, 1577; Dutch mountain section, 1582; Mehoop-
    any well section, 1582; Tioga river section, 1584;
    Terrace, 1572, 1659; formation, 1781, 1783; mountain
    plateau, 1651, 1568. 1437; Allegheny escarpment,
    1573; in Western Pennsylvania, 1573; in New York.
    1829, 2153; around Broad Top, 1574; in Schuylkill,
    Carbon and Monroe counties, 1574; Little mount-
    ain, 1574; in Fulton and Bedford counties, 1613;
    Yellow creek section, 1615; Hyndman section, 1615;
    along Allegheny mountain. 1616; in Clinton county,
    1617; in Lycoming county. 1621; at Honesdale, 1585;
    Dyeberry creek well, 1587; why no coal basins in,
    east of Lehigh, 1589; Bluestone quarries on the
    Delaware, 1590; Bluestone quarries in Wyoming
    county, 1591; Newport limestone in Perry county,
    1594; on the Lehigh river. 1594; Coxton section,
    1607; Campbell's ledge, 1607, around Broad Top
    mountain. 1609; Hann's Ridge group, 1611; Sideling
    Hill, 1612; fossils, Chapter CX, 1623; fossils on the
    Susquehanna and Juniata river. Chapter CIX, 1601.
    in Perry county, 1601; iron ore in Perry county,
    1603; on North branch of Susquehanna river, 1604;
    sandstone and shales No. IX, ...1643,1653,2220
" and Chemung in four Ohlen wells in Potter county, Chapter
    XCIX, 1471; limit. Chapter XCV, ...1403
" " Pocono formations, 1641; wall, 1254; plateau, 1571; transi-
    tion beds IX-X, ...1643
Cat's run mine, Pittsburgh coal at, ...2560
Cauffield's mine, Pittsburgh coal at, ...2472
Caves, precipitation of limonite, ... 433

Caverns in No. II, Chapter XXXVI, ... 425
Cayuga colliery, ... 1972
Cedar Hill colliery, 2112; mine, ... 2521
Cement, Johnstown, ... 2307
" quarries and paint mines of the L high Water Gap, ... 1067
Cemetery anticlinal, ... 1986
" Hill section, ... 2571
Central colliery, ... 1978
Centralia basin and colliery, ... 2063
" drainage tunnel, ... 2060,2063,2064
" -Mt. Carmel division, ... 2062
Centre county No. XI in, ... 1840
" " anticlinals, 365; Gatesburg ridge anticlinal, ... 367
" " barren measures in, ... 2410
" " fault and crush in, Hamilton No. VIII rocks, ... 1218
" " geology, ... 500
" " limonite mines, Chapter XXXII, ... 372
" " measurement of No. IV, ... 631
" " No. VI in, ... 999,1002
" " No. VII in, ... 1103,1120,1123
" " No. XII in, ... 1865,1867
" " No. XIII in, ... 2194
" " Portage in, ... 1355
" " Sand ridge (Tadpole ridge) anticlinal in, ... 367
" " (white limestones and marbles of N. II in Chester, Montgomery, York and Centre counties) Chapter XXXIX. 467
Centre, Clinton and Lycoming counties, No. V in, ... 858
Centreville mines, Elk county, No. XIII rocks, ... 2285
Ceylon opening, Waynesburg coal in Pool No. 6, ... 2561
Challenger pit, Pittsburgh coal, ... 2541
Chamberlain colliery, Southern coal field, ... 2114
Chambersburg and Bedford turnpike, 996, 998, section along, ... 840
Champion mine, Pittsburgh coal, Washington county, ... 2554
Champlain Dist., Dr. Emmons' survey of the, ... 40
" lake, ... 158,165
Chapman division (slate), Bangor slate belt No. 111. 583
" shaft, Northern coal field, ... 1978
Chatham anticlinal, ... 2276
Chatham-Farmingham anticlinal, ... 1436
Chance, H. M. measurements at the Delaware and Lehigh, Schuylkill and Susquehanna river gaps, ... 641
" " measurements at Mill Hall gap, Clinton county, ... 647
" " measurement of No. IV in Clinton county, ... 631
" " measurements of No. V at the Delaware Water Gap, 730
" " measure of No. V at Lehigh Water Gap, ... 731

Chance's long section Rep. G4, fish beds, ...1463
" map of the Delaware Water Gap, 1055; section of Broadhead's creek gap, ...1055
" Dr., topographical map and long section, Lehigh river, ...1063
" vertical cross section at Schuylkill Water Gap, ...733
Character of formation No. III, Chapter XLVII, ...562
Charles, Abraham, copper ore indications, ...2632
Charles Pott coal bed, Southern coal field, ...2103,2114
Chauncey colliery, Northern coal field, ...1990,2004
Cheat river gap, ...1846
Checker coal bed, Northern coal field, ...1981
Chemung No. VIIIg, Chapter XCIII, 1364; in Middle Pennsylvania, Chapter CV, 1546; in Northern Pennsylvania, Chapter XCVI, 1432; in Susquehanna county, 1439; in Bradford county, 1443; in Erie county, Chapter CII, 1497; sections, 1378; stratification, 1380; mineral, 1381; Catskill in four Ohlen wells in Potter county, Chapter XCIX, 1471; oil sands, Chapter C, 1481; conglomerate, 1549; fish beds, Chapter XCVII, 1453; fossils, Chapter XCVIII, 1468, 1373, 1383, 1561; upper limit of, Chapter XCV, of New York, Chapter XCIV, 1779; valley, ...1723
Cherry Grove (oil field), ...1491
Cherry Tree well record, ...1870
Chester county, copper operations in, ...355
" " magnetic limonite mines, ...256,262
" " map, ...127,202,203
" " reports, ...44,91
" " serpentines, ...103
" " Warwick and Pickering Valley ores, ...268
" " White limestones and marbles of No. II in Montgomery, York and Centre counties, Chapter XXXIX, ...467
Chester Wilson slope, Broad Top coal field, ...2172
Chester Valley marble quarries, ...468,110
Chestnut Ridge, ...1699,1701,1709,1789,1807,1839,1843,1844,1845,1846,1891,2160,2473
" " anticlinal, ...367,1225,1116,1861
" " axis, ...1889,2303,2316,2321,2323,2330
" " gap, ...1797,2329
" Hill group, Philadelphia gneiss belt, ...125,123
Chrisman, F. coal opening, Mehoopany coal basins, ...2013
Church coal bed in Heckscherville basin, ...2085
" colliery, Northern coal field, ...1972
" drift coal bed, Northern coal field, ...2005
. " slope coal bed. Northern coal field, ...1973

Cincinnati, 27, 535, 718, 1345; anticlinal, 1037; country, 621; rocks, 1033; region, ... 633
Clarion bed, coal A', Elk county district, ...2282
Clarion-Butler belt, ...1543
Clarion coal bed A', ...2238,2294,2313,2317,2390
" group, description of, ...2214,2368,2377,2383,2395
" county No. XII in, 1895; No. XIII in, ...2373
" oil belt, 1490; oil field report, Carll, ...1501
" series, beds A and A', ...2357
" upper coal, ...2379,2393
" -Venango oil fields, ...1544
Clarissa mine, Pittsburgh coal, ...2491
Clark, A. B. mine, Indiana county, Kittanning group, ...2325,2327
" bank Jefferson county, ...2307
" bed, Northern coal field, ...1970-79
" and McCormick's drifts, Tremont, Southern coal field, ...2124
Clarkson coal bed in Southern field, ...2103,2114
Clarkville mine, Waynesburg coal, ...2561
Claypole, list of Perry county plants in F2, ...1270
Claysville axis, ...2457
Clearfield coal districts, ... 372
" coals, analyses of, ...2465
" coal mines, ... 295
" fire clay, analyses of, ...2218
" county, ...1840
" " barren measures in, ...2410
" " No. XII in, ...1865,1868
" " No. XIII in, ...2205
" and Cambria region, ... 999
Clear Spring colliery, Northern coal field, ...1981,1982
Clepsysaurus Pennsylvanicus, ...2608
Clermont basin, ...1876,2278,2289
" " vertical section in, ...1878
" coal (Clarion A' bed), ...1876,1880,2278,2280,2282
Clifford colliery, Clifford coal bed, Northern coal field, ...1956.1957
Cliff mine, Pittsburgh coal, ...2544,2552
Climax mine, Pittsburgh coal, ...2521
Clinton coal bed, Southern coal field, ...2114
" county, ...1833,1840
" " general section of strata, ... 840
" " geology, ... 500
" " measurement of IV in (Chance), ... 631
" " No. VI in, ...999,1003
" " No. IX Catskill in, ...1617
" " No. XI in, ...1837
" " No. XII in, ...1865,1866

Clinton county, No. XIII in,	2190
" " Portage,	1356
" " Rep. G4,	647
" Centre and Lycoming counties, No. IV in,	667
" No. V formation, Chapter LV to LXIII, 721; Great thickness of, 750; Upper Olive shales, 757; Middle Olive shales, 814; Lower Olive shales, 815; Danville ore, 758; lower lime shales, 758; upper lime shales, 759; ore sand rock, 758; sand vein ore bed, 758; fossils, 759; in Huntingdon Valley, 825; along Tussey mountain, 836 to 837; in Bedford county,	840
Clipper mine, Pittsburgh coal,	2554
Coal beds, different names of, 1924; correlation of,	1951
" breakers of iron,	1934
" Brook colliery,	1955,1957
" " slope,	1957
" false beds, Marcellus,	1217
" Glen mine, 2298; colliery,	2306
" Hill colliery,	2097,2100,2101
" measures in the anthracite region,	1923,1924
" " in the Eastern Middle field,	2024
" " in Northern anthracite field, 1951,1954,1960,1968,1976,1986,2002	
" " of the Southern field,	2075,2087,2107,2139
" " in the Western Middle field,	2051,2055,2059,2064,2068
" none of value in the New Red formation,	2628
" places in Montgomery county,	2629,2630
" vegetable origin of,	1929
" washers,	1933
" Waste Commission Report, Reports A2 and AC.,	1933,1935,2147
Coast Survey soundings, 18, report,	1882,273
Coburn well, Chemung rocks, McKean county, 1487; Pocono S. S. in...	1733
Cobus opening, Lower Freeport coal D, Indiana county,	2332
Cochran opening, Pittsburgh coal,	2490
Codorus limestone beds,	454
" ore bank, 253; Codorus furnace, 253; P. O.,	223
Colebrook furnaces,	317
Coleman heirs' quarries,	315,316,317
" opening, Berlin coal bed of No. XIV,	2421
" pit, Pittsburgh coal,	2529
Colket colliery,	2120,2121,2122,2123
College and Pennsylvania furnace group,	378
Collieries, increased capacity of,	1934
Collier's (W.) flagstone quarry,	615
Colorado, Grand Canon of the Arkansas river,	756
" 425, 503, 1627; well, 1491; No. 4 (well), 1543; well record,	1539
Color of the rock beds and aid to their sub-division, New Red,	2606
Columbia, Chicques rock at,	40

INDEX FINAL SUMMARY REPORT. 23

Columbia mine, bed D, Cambria county, 2236; Pittsburgh bed, ...2518
 " county, 27, 737, 781, 821, 885, 918, 939, 947, 950, 960, 962, 993, 1009,
   1022, 1025, 1031, 1039, 1082, 1133, 1222, 1259, 1261, 1329,
   1315, 1318, 1364, 1409, 1410, 1425, 1428, 1467, 1551, 1552,
   1555, 1563, 1564, 1570, 1571, 1601.
 "   " quarries, ...948,949,960
 "   " White's section in, ...1329
Columnar section, general, combined from certain parts of the field,
  New Red, ...2604
 "   " sheet lists of, Northern coal field, ...1941,1942,1943,1945
 "   " sheets, plan of, anthracite district, ...1936
 "   " combined into a general one, New Red, ...2605
Comer's quarry, Venango second oil sand, Erie county, ...1510
Condition of the coal beds in Western Middle coal field, ...2051
Conemaugh gap, ...1709,1721,1723,1795,1842,1870
 "   " sections in, ...1703,1705,1799,1844,1849
 " river, 1451, 1573; gap, ...1247,1413,1500,1511
Conestoga valley, 228, 268, 270, 289, 487; limestone, ... 153
Confer anticlinal, ... 369
Conglomerate of "Beaver River Series," general section. 1911
 " No. XII, ...1841,1860,1888,2153,2237,2248,2253,2264
Conneaut valley, ...1498
Connecticut mine, Pittsburgh coal, ...2519
 " New Red, ...2600,2611,2614,2615,2616
Connedogwinit creek, ...272,287,297
Connellsville (Blairsville) basin, ...1850
 " 1573; coal district, 372; Mountain gap, 1709 and, ... 28
 " coke production, tables of. 2478,2479
 " coking basin, ...2472
 " region, list of coke ovens in, ...2479,2480
Connor, G. W. mine, Waynesburg coal, ...2561
 " mine; Alton coal group and tract, ...1883
Connoquenessing sandstone, ...1785,1863,1888,1897,1911,1913,2163
 " lower sandstone, 1906; upper sandstone, ...1905
Conococheague creek, 143, 149, 150, 272; valley, 149; backset, 145; first
  head brook of, ... 247
Conshohocken, 79, 83, 93, 128, 166, 174, 211, 329, 469, 477, 505; axis, 468;
   dyke, 452, 454; mines, 114; furnace quarries at West
   Conshohocken. 81
 " trap, ...2620,2625
Consolidated colliery, Northern coal field, ...1978,1980
Continental colliery, Northern coal field, ...1970
 " quarry, No. III rocks, ... 601
Conyngham colliery, Northern coal field, ...1987,1991,1996,1997
Cook bed, Broad Top coal field, ...2174
 " pit, Greensburg basin, Pittsburgh coal, ...2500

Cooley bank, Pittsburgh coal, ...2539
Cooper coal bed, Northern coal field, ...1991,2006
Coopersburg trap, ...2636
Copper in New Red, ...2631,2632
Corbin colliery, Western Middle coal field, ...2070
Corniferous No. VIIIa limestone, 1153; in Pennsylvania Chapter LXXXIII, 1170; on the Lehigh, 1173; on the Delaware, 1170; in Western Pennsylvania, 1189; in Perry county, 1182; as a source of petroleum, 1191; fossils, ...1192
Cornwall iron mines, 309, 354; magnetic iron ore mine, ...304,846,1387
" mine, ...256,257,268,344,351,355,356
" trap dyke, 309; fault, ...309
Cornell and Werling mine, Pittsburgh coal, ...2533
Corry sandstone, ...1769,1773
Cortney's bank, Pittsburgh coal, ...2539
Couch mine, Pittsburgh coal, ...2556
Coudersport basin, ...1872,2277
" 1463, 1464, 1465, 1471; Mt., 1435; quarry, 1464; synclinal, ...1875,1464,1478
Courtney's mine, Clarion coal, Mercer county, 2390; Pittsburgh coal,..2552
Cove anticlinal, ...1281
" mountain, ...1651,1655
" " estimate of No. IV in, ...631
" synclinal, ...643
Cover bank, coal E, Somerset county, ...2247
Covode openings, coal E, Somerset county, ...2260
Cowan mine, coal C, Jefferson county, ...2309
Cowanesque basin and synclinal, ...1478,2276
Coxe, Eckley B. Coxe, Bros. & Co.,......................1928,1933,1934,2039
Coxton section, ...1643
Crabtree mines, P. C. Greensburg basin, ...2498,2500
Crawford coal works, ...1908
" county map, ...1492,1506
" " No. XII in, ...1893
" " Rep., 1430; Rep. G4, 1367; Rep. Q4, ...,.........1497,1584
Creek opening, Clermont coal, Cameron county, ...2281
Cresson, Dr. Charles M. analyses of Leh. C. & N. Co. coals,...........1928
" shaft, section of No. XIV, bed E at, ...2412
Crinoidal limestone, ...2408
Crooked creek mountain, ...1723
" " 1112; valley, 1277; synclinal, ...1459,1460
Crosby coal bed, Southern coal field, ...2094
Cross-bedding theory, New Red, ...2594

Cross creek coal basin,	2028
" " opening, Pittsburgh coal,	2510
" Cut coal bed, Southern coal field,	2090
" section sheets, list of,	1941,1942,1943,1944
" section sheets, plan of,	1936
Crouch's pit, Pittsburgh coal,	2557
Crowl mine, bed C', Somerset county,	2261
Crow's Ferry opening, Pittsburgh coal,	2563
Croyle bank, bed E, Somerset county,	2259
Crystalline rocks of the Northwest, N. H. Winchell,	156
Culm banks, coal in,	1933
" to fill in old mine workings,	1934
Cumberland basin, analyses of Pittsburgh coal bed in,	2465
" Co. Geol. map, 324; colored map, Atlas D5,	455
" county limonite mines,	231,238
" " S. Mtns. of, 62, 70; Dr. Frazer's section of in S. Mtns.,	45
" " trap dyke in,	456,458,564
" Md., 26, 27, 756, 1000, 1030, 1135; coal basin,	681
" mine, Kelly coal in, Broad Top basin,	2171
" mountain,	2154
" plateau,	2154
" Valley dyke,	458,460
" " report on the iron ores, etc., of, by E. V. d'Invilliers,	238
" and Elk Lick mine, Pittsburgh coal,	2462
Cummings bank, bed E, Indiana county,	2339
Cunard mine, Kelly coal section in, Broad Top basin,	2170
" shaft, Broad Top basin,	2173
Curfman opening, Broad Top basin,	2181
Curry pit, Pittsburgh coal,	2531
" -Woodbury anticlinal,	409
Cussewago limestone,	1773
" sandstone,	1647,1775,1777
" shales,	1773
Custer bank, bed C', Somerset county,	2251
Cymbria mine, bed D, Cambria county,	2240

D.

Daddows' coal opening, Wyoming county, 2014; old slope,	2113
Dagus (Lower Kittanning) bed B,	1876,1878,2278,2281,2284
" mines, Elk county,	2285
Daniel Beer's quarry (slate),	596
" coal bed, Southern coal field,	2034
" Webster colliery,	2071
Daniel's quarry (Slate No. III),	596

Danville bed, 803; ore bed, 811, 812; in Huntingdon valley, 827; fossil
　　　ore, ...813,853
　" 　gap, 737, 739; section 747,748
　" 　sections near, 748; section below, ... 891
　" 　workings, ... 752
Dark Hollow, ...148,1268
Darlington coal bed C', ...2160,2392,2394
　" 　shales of Beaver county (Euripterids in), ... 9
Dark Hollow coal, New Red, ...2629
Dauphin county, ...1861
　" 　　" 　coal basin, ...169,643,691,705,1130
　" 　　" 　colored map, ... 452
　" 　　" 　limonites, ... 231
　" 　　" 　No. VI in, Chapter LXVII, ... 936
　" 　　" 　quarries (limestone of No. II), ...311,319
　" 　　" 　Sanders' dip map of, ... 322
David (James), Hyner's run oil well record, ...1621
Davidson's quarry, No. III slate, 595; sandstone No. VIIIg, Erie
　　　county, ...1353
Davis bank, bed B, section at, ...2215
　" 　(Mine No. 1) mine, limonite ores, ... 406
　" 　opening, Pittsburgh coal at, ...2485
　" 　bed D, section at, ...2216,2217
Dawson mine, bed E, Upper Freeport coal at, ...2404
Dean mine, Pittsburgh coal at, ...2521
Dean's bank, limonite ore, ...402,404
Deener's bank, Kittanning, middle coal C at, ...2387
Deep bank, Lower Freeport coal D at (Depp bank), ...2324
　" 　well, Warren county oil district, ...1509
Deer Lick opening, Clermont coal A', McKean county, ...2279
Decker Valley (Confer anticlinal in No. III), ... 369
Decker's Ferry, analyses of Stormville cement bed, ...924,927
Deckertown pike, slate, analysis of, ...564,567
Deckert's quarry, limestone No. II, Berks county, ... 311
Deck's (James) quarry, No. III slate, ... 595
Delabole quarry, No. III slate, ... 594
Delano-Shenandoah division of Mahanoy basin, Western Middle coal
　　　field, ...2052
Delaware county, Kennedy's granite quarry, 97; map, 127, 128; Rep.
　　　(by C. E. Hall, 44, Rep. C4, 50; (Serpentine), 102; gap, roof-
　　　ing slate, 564, 567; section, 1067; and Hudson Coal Com-
　　　pany's No. 4 shaft, 1988; Plymouth collieries, 1993, 1996;
　　　River Bend, at Port Jervis, 773; in northern point of New
　　　Jersey, 905; below Trenton, 49; river, blue-stone quarries
　　　on the, 1590; buried channel, 1203; old channel, 1248; divide
　　　between, and Susquehanna, 1582; generalized section of No.

VI on the, 918, 920, 923; No. VII on the, and Lehigh rivers, Chapter LXXVI, 1045; Portage rocks on the, 1361; quarries on, 308; section, 1428, 1584, 1585, 1587, 1590; (section along eastern bank), 575; sections between, and Lehigh, 1598; (slate beds), 550; the Corniferous on the, 1170; Walpack bend, 1047, 1048, 1049, 1051; cross section of Hogback, 1049; and Schuylkill series of New Red beds, 2603; section, Chemung rocks of the, 1426; Water Gap, Dr. Chance's map of, 1055; (measurements at), 641; measurement of No. V, 730; No. IV at the, 675; No. VI west of the, 920; old quarry, ...583
Dellville, 1558, 1563, 1603; road, 1268; sections, etc., ...1558,1559
Delp property (old), quarry, East Bangor slate No. III, ... 584
Delta mine, bed D at, ...2240
Delthyrus limestone of No. VI, ... 906
Dennis' (Adam) tannery bore hole, ...1452
Dennis run (well record), ...1539
" well, ...1465,1533,1535,1536
" No. 1, 1482, 1483, 1484, 1487; section 1533,1534
Deppen's (Samuel) quarry, No. II limestone, ... 313
Depth of mine workings, Northern coal field, ...1985
Derringer colliery, Eastern Middle coal field, ...2036
Derr's quarry, No. VI limestone; sponge corals, ...1025
Derry, 159, 1029; church, 319; quarries, 320; station, ... 319
" Coal and Coke Company's mine, ...2483
Deshong bank, Cambria county, ...2238
Detweiler's quarry, conglomerate limestone at, ... 473
Devil's Den, ...1307
" well, 1061; Delaware Water Gap, Lehigh Water Gap, ...1055
Devonian and Carboniferous systems, ...1641
" formation, 1821; lower limit of, ... 889
Devores quarry, sponge corals of No. VI, ...1029
Dewees, Royer and, shaft and tunnel, Marcellus ore No. VIII, ...1230
DeWitt's (J.) section, Stormville shales, ... 929
Diabase, character and occurrences, New Red, ...2624
Diamond coal bed in Northern coal field. 1965,1972,2007
" " " in Southern field, ...2102,2113,2123,2132
" " " in Western Middle field, ...2057,2062,2071
" colliery, Northern coal field, ...1970,1971,1972
" " Southern coal field, ...2123
" " Western Middle coal field, ...2061
" mine, bed D, Jefferson county, ...2305
" slate quarry, roofing slates of No. III, ...508,608
Dick's ridge, 1211, 1269; gap of Juniata river through, ...1268
Dictionary of Fossils, Pa. Rep. P. 4. J. P. Lesley.
140,190,497,892,893,916,1009,1018.1167,1333,1349,1428
Dill, J. C. mine, bed E, Indiana county, ...2320

Dillner's pit, Pittsburgh coal, Greene county, ...2563
D'Invilliers, E. V. Ann. Rep. 1886, ...238,240,311,319,326.358
" topographical map of Cornwall mines, ... 357
" F. V., prefatory letter of, Vol. III, Part I, ...1855,1856
" Report Berks county, 66; Rep. D3, ...235,267,269
" Rep. F3 (Union and Snyder counties),
421,422,983,991,1077,1078,1097,1276
" Report T3, Huntingdon county, ...367
" Report T4, Centre county, ...165,367,365,372,1218
Dimension, geological, Chapter III, ... 38
Dips described, New Red district, Bucks and Montgomery counties,...2598
Doan's lead mine, New Red, ...2631
Dobbin's quarry, Corry sandstone, ...1771
" well, third sand in, ...1507
Dochranamom hill, Potsdam (Primal) sandstone in, ... 201
Dochterman's ore bank, Hamilton formation, ...1268
Docker and Bowman's slope, Southern coal field, ...2103,2104
Doddridge quarry, No. III roofing slate at, ... 602
Dodson colliery, Northern coal field, ...1993,1994,1995,1996,1997
Dorbin's slope, Southern coal field, ...2121
Dolerite, character and occurrences of, New Red, ...2624
Dolph colliery, Northern coal field, ...1961
Donges' quarry, Myerstown, limestone of No. II, ...310,314
Donnelly's bank, Pittsburgh coal, ...2485
Doolittle's quarry, red shale, Erie county, section, etc., ..1510,1513,1514,1517
Dorrance colliery, Northern coal field, ...1993,1996,1997,1998
Dorsey banks, limonite ore in Sinking Valley, ... 398
" ore deposits, ore in Sinking Valley, ... 367
Dotterer red hematite mine, Berks county, ...268,269
Doty mine, Pittsburgh coal, Indiana county, ...2467
Dougherty bank, Cook bed, East Broad Top, ...2181,2182
Douglass bank, limonite of No. II, 252; Marcellus ore, ...1231
" slate quarry, No. III formation, Douglassville, ... 596
Dowler coal bank, bed B, Clearfield county, ...2215
Doylestown, 84, 179, 452, 505; fault, ... 555
Drake's quarries, Connoquenessing SS., Mercer county, ...1905
Dreck creek basin, Eastern Middle coal field, ...2041
Drexel's quarry, limestone of No. II, Berks county, ... 311
Drifton colliery, sp. gr. of coal at, 1929; Eastern Middle coal field, 2029,2030
Driftwood or Second anticlinal axis, ...2190,2280
Dry Hollow bank, limonite ore of No. II, ... 394
" " synclinal, Centre county, position of, ... 367
" run mine, Pittsburgh coal, ...2552
DuBois basin, ...2303
Duck & Co., Young quarry, slate of No. III, ... 595
Dudley basin, Broad Top coal field, ...2177,2409

Dull and Bradley ore bank, Oriskany ss., ...1226,1227
"   "   "   sand mine, Oriskany ss., ...1035
"   "   " Hoff lands, coal beds on, ...2142
Dunbar furnace, ...1844
"   section, ...2435,2493
Duncannon (West and East) dykes, ...458,460,462,463
Dundas tunnel, in Sharp mountain, Southern coal field, ...2120,2121
Dundee shaft, coal measures at, Northern coal field, ...2002
Dunkard coal bed, Greene county, No. XVI, ... 2582
"   creek region wells, ...1543
Dunlap bank bed D, section at, ...2213
Dunmore anticlinal overlap, ...1968
"   coal beds of Jermyn—Priceville, ...1961
"   No. 1 coal bed, ...1969
"   No. 2, coal bed, ...1969
"   No. 3 coal bed, ...1968
Dunn colliery, Northern coal field, ...1980
Dunning's Narrows section, ...840,846
Dupont's drift, Northern coal field, ...2003
Duquesne and Hampton mine, Pittsburgh coal, ...2531
"   mine, Pittsburgh coal, ...2528
Durham hills, 70, 161, 179, 704; and Reading hills, 61, 74; Highlands, .. 351
Dutch Corner, 684, 728, 843, 849; cove, ... 850
"   mountain, White's section, No. VIII, IX and X, ...1643
Duval mine, Kelly coal section in, Broad Top basin, ...2170
Dyeberry creek, 1580, 1585; well record, ...1587
Dyer's run, shaftings at, Southern coal field, ...2081
Dysart's mine, ore sandstone of No. V, ... 854

E.

Eagle colliery, Southern coal field, ...2105,2108,2109,2111
"   Hill colliery, Southern coal field, ...2105,2109,2110,2112
"   mine, Pittsburgh coal, ...2537
"   slate quarry, roofing slate of No. III, ... 600
Eagleton-Furney axis, ...2190
"   coal mines, Clinton county, ...2190
Ealy's quarry, Girard shale at, ...1354,1368
Earliest fish known, 756; England, ... 769
East Allen, Northampton county, quarries, roofing slate, 599; Bangor quarries, roofing slate of No. III, 590; Boston colliery, Northern coal field, 1988, 1990, 1992, 1994, 1995; Broad Top coal field, 2180; Clarion spring water, Elk county, analysis of, 448, 449; Duncannon dyke, 458, 461, 463; Franklin-Brookside division, Southern coal field, 2126; Franklin colliery, 2122, 2123, 2127, 2130, 2131, 2132; Mahanoy railroad tunnel, 2051, 2055; Stroudsburg, 1171, 1173; anticlinal, 1173, 1174; Tennessee, the Great Valley, geology of, 500; Whiteland (Trimble's iron mine), ... 114

Eastern Middle anthracite field, described, ...2022
"   "   "   " location of, ...1918
"   " coal field, contents of, ...2148
"   " field, Pottsville conglomerate in, ...1854
Easthampton, Mass., fossils, ...2614
Easton-Reading or Durham hills, ... 70
Eaton colliery, Northern coal field, ...1963
Eberhard's quarry, magnesian limestone beds in, ... 329
Ebervale colliery, Eastern Middle coal field, ...2033
Ebony colliery, Southern coal field, ...2083
Echo lake, 1203, 1206; quarry south of, ...1172
Eckel's tunnel, Southern coal field, ...2124,2125
Eckert colliery, Southern coal field, ...2130
" (John), analyses Portland cement, ... 338
Eckert's quarry, Lebanon Valley limestone, ... 314
Eckley collieries, Eastern Middle coal field, ...2033
" overturned anticlinal, ...2032
Ecks' quarry, 953; section, ... 959
Eclipse mine, Pittsburgh coal, ...2554
Economic geology of the New Red rocks, ...2626
Ecton mines, copper ore in the New Red, ...2631
Eddy Creek colliery, Northern coal field, ...1965,1973
Edgerton colliery, Northern coal field, ...1963
Eisenhuth Run anticlinal, ...2078
Eldersville opening, Pittsburgh coal, ...2540
Eleanora mine, bed D, Jefferson county, ...2240,2241,2298,2302
" mines Nos. 1 and 2, bed D, Lower Freeport coal. 2297
Eleven Foot coal bed, Northern coal field, ...1990
Elizabeth mine, limonite ore of No. II, ...1002
" copper mine, Knauertown or French creek, ... 267
Elk county, No. XII in, 1858, 1866, 1881; county, No. XIII in, 2279, 2281; county, 1487; oil borings, 448; oil fields, 1485; oil sands, 1486; oil sand group, 1485; Lick coal, Barren measures No. XIV, 2407, 2415; mountain, 1651, 1721; mountain range, ...1647
Elko pit, Pittsburgh coal, ...2529
Ellengowan colliery, ...2058
Ellenville, 635, 1201, 1478; mine of lead ore in No. IV, ... 678
Ellicottsville conglomerate, ...1521
Elliott and Bryson bank, bed B, Clearfield county, ...2214
Elmwood colliery, Northern coal field, ...1978
Ely and Riehle tracts, Southern coal field, ...2079
Emaus, 180, 254; Iron Company mine, ... 234
Emerich and Lebo's quarries, 963, 964; section, ...1083
Emery mine, Waynesburg coal at, ...2556
Emmett mine, Pittsburgh coal at, ...2528
Empire colliery, Northern coal field, ...1987,1989,1992,1994,1996

Empire quarry, No. III slate, ... 597
Emporium, 1356; well and section, ... 1488
England (Wenlock formation), ... 879,887,1032
" Wigan coal measures, 188; coal measures of, ... 190
Engle's quarry, Dauphin county limestone No. II, ... 321
English geology, 20; systems of the English geologists, ... 41
" mine, bed B in, ... 2189
" nomenclature, 901; geological names, ... 46,48
" and French Portland cements, ... 339
Ennis Hill, conglomerate series in, ... 1895
Ennisville anticlinal, ... 1109
Enterprise colliery, Northern coal field, ... 1990,1995,1996
" " Western Middle coal field, ... 2069,2070
" mines, Pittsburgh coal at, ... 2545
" sand works, No. VII in Juniata valley, ... 1096
" well, oil region of Warren county. 1491
Ephrata, quarry at, marble of Lancaster county, ... 481
Epidiorite, character and occurrences of, in New Red, ... 2625
Erb's quarry, Dauphin county limestone of No. III, ... 320
Erie colliery, Northern coal field, ... 1961,1963
" county, Chemung in, 1496, 1497 to 1500; map, 1506, 1492; general
section in, 1497; Girard shale of, 1367 to 1370; outcrop Venango
oil group, 1501; petroleum in, 3d sand in, 1517; VIIIf Portage in,
1351; Prof. White's section of Chemung in, 1492; quarries, 1512,
1517; Rep., 1430; G4, 1367; Q4 White, 1584; Venango 1st oil sand
in, 1507; Venango 2d oil sand in, 1509; Venango 3d oil sand in,
1511; Venango lower shales in, 1510; Venango upper shales in,.1509
" well Trenton limestone in, 496; Corniferous in. 1190
" (wells at), 26; Presque Isle well at, ... 27
Erosion, rate of, ... 490
Eskdale, coal measures in Scotland, ... 1017
Etna bank, limestone and ore, ... 402,403
Eureka colliery, Western Middle coal field, ... 2071
" oil well, ... 1526,1527
" section, Olenellus found, ... 192
" tunnels, Southern coal field, ... 2120
Europe (glacial period), ... 51
European sections, ... 1311
Euryptirids in the Darlington shales of Beaver county, ... 9
Evans colliery, Eastern Middle coal field, ... 2043
" No. 21 well, ... 1543
" pit, Pittsburgh coal. 2523
" quarry, 311, 950; quarry No. 1, 952; quarry No. 2, 953; abandoned
quarry, ... 1025
Evarts upper tunnel, ... 2123
Ewing mine, Panhandle district, Pittsburgh coal. 2541

Ewing H., mine, Pittsburgh coal, ...2468,2470
" Prof. special Rep. in T4, ...425,479,1121,1122
Exeter colliery, Northern coal field, ...1981,1982
Experiment Mills limestone quarry, ...929,933,934
Export mine, Pittsburgh coal, Westmoreland county, ...2508
Eyer's bank, Nittany Valley ore mine, ... 395

F.

Facundus well record. 1537,1539
Fairbanks (McGregor) mine, bed D, Indiana county, ...2342
Fairview well section, mountain limestone in, ...1715
Fairhaven and Paultney, Vt. (slate belt), description of, ... 583
Fairlawn colliery, Northern coal field, ...1969
Fairmont colliery, Northern coal field, ...1978
Fairmount basin, Mercer coal group in, ...1887
" Coal Company's No. 1 opening, bed D in, ...2160,2376
Fairplay basin, Broad Top coal field, ...2172
Falkenstine shaft, brown hematite of York county in, ... 222
False-bedding theory, New Red formation, ...2594
Fall Brook Coal Company's mines, description of, ...2270
Falls creek, 1370, 1371, 1441, 1553; tunnel on, ...1371
" " axis, ...2302,2305
" of Schuylkill, ...120,121
" of the Ohio, ...621,1192
" Run basin, Eastern Middle coal field, ...2035
Fannettsburg ore bank in Path Valley, fault at, ... 252
Farmington belt, Chemung rocks of Tioga county, ...1436
" wells, Venango oil group in, ...1543
Farrandsville, No. XII at, 1866; long section No. IV to No. XIII,..1619,1620
Fathomless Spring, source of Penn creek, Centre county, ... 423
Faulkner shaft, description of limonite ore in, ... 412
Fault in Perry county, 979; and crush in Marcellus, Centre county, ..1218
Faust bank, bed D, section at, ...2214
" coal bed, Southern coal field, ...2116
Faust's glass sand quarry, ...1103,1105
Fayette county, ...1852
" " No. XII in, 1891; No. XIII in, ...2362
Featherman's quarry, Bossardville limestone of No. VI at, ... 932
Federal Spring mine, Pittsburgh coal at, ...2541,2542
Feeder Dam colliery, Southern coal field, ...2113
Feger Ridge colliery, Southern coal field, ...2125
Feisterville, Van Arsdale quarry gypsum crystals, 450; graphite, ... 478
Felty's drift, Southern coal field, ...2123
Fell township, Huntingdon county, Oriskany S. S. described, ...1090
Ferndale colliery, Western Middle coal field, ...2065

Fernwood anticlinal, ... 1975
" shaft, Northern coal field, ... 1973
Ferriferous limestone,
2161,2281,2292,2294,2310,2328,2344,2352,2365,2378,2380,2387,2389,2401
Fifth bituminous coal basin, Alton group in, 1833; Clermont coal, 2278;
description, ... 2288
Fillmore-Boalsburg cross section of No. II limestone, ... 371
" -Reynoldsville basin, Lisbon synclinal, ... 2295,2302
Findlay opening, bed B at, Blairsville basin, Indiana county, ... 2335
Findleyville mine, Pittsburgh coal bed in, ... 2561
Fire clay analyses in Centre county, ... 2202
First Anthracite coal field, southern border of, 705; bituminous basin along Beech creek, 2199; bed B in, 2250; coal bed, Lance coal, Northern coal field, 1994; or Laurel Hill axis, Clearfield county, 2205; limestone, Washington county group, No. XVI, 2578; (Twin) coal bed, Southern coal field, 2093; (upper red ash) coal bed, Southern coal field, ... 2092
Fish beds in VII and IX, Chapter LXXXXVII, ... 1453
" earliest known, 756; of England, ... 769-772
" creek sandstone, ... 2582
Fisher colliery, Barnet coal in, Broad Top coal basin, ... 2177
Fisher's quarries, No. VI limestone in, ... 1002
Fishing Creek gap, ... 742,747,1093
" Little, creek, No. IV in, ... 370
" creek section, 743, 944, 1073, 1555, 1320; No. VI section, ... 942
" (Big) creek section, ... 1266
" (Little) creek section, ... 1266
" creek valley, trap dyke in, ... 456
Fishkill, Valley of the, ... 1195
Fishpot limestone, ... 2451,2471,2474,2503
Fitzpatrick's colliery, Southern coal field, ... 2100
Fitzwater, Mr. quarry on farm of, marble, ... 470
Five Foot coal bed, Northern coal field, ... 1994,1995
Flat Brook Valley, N. J., No. VI in, ... 920
" Top coal series in West Virginia, ... 1858
Fleck farm, 445; pits, zinc-lead deposits in No. II, ... 447
" (old) open cut, specimens of Marcellus ore, analyses, ... 1230
Fleming's gap, Clinton Olive slate in, ... 863
Flora of the coal period, ... 1930
Florence bank, Pittsburgh coal, ... 2540
Flynn quarry, roofing slate of No. III in, slaty foliation, ... 579
Fogarty colliery, Southern coal field, ... 2116
Fontaine's determination of the Rhaetic age of Virginia and North Carolina fossils, ... 2611
" "Kaolin slate" beds in Virginia Blue Ridge, ... 144

Foot-prints of shore feeding animals, ... 17
" from Silliman's Journal, ... 773
Forest county, 1485, 1489, 1502, 1305; wells, 1489; oil sands, ...1486
" City-Carbondale division of Northern coal field, description of, 1952
" " colliery, Northern coal field, ...1957
" " description of coal measures about, ...1952
" " identity of coal beds at, ...1955
" county, No. XII in, ...1881,1885,1886
" " No. XIII in, ...2279,2291
" City No. 2 shaft, Northern coal field, ...1955,1956,1957
Forrestville colliery, Southern coal field, ...2118,2120,2121,2122
Forge or E coal bed, ...2096
Forksville deposit, coal in, Mercer lands, ...2184
Formations, thereories of, 723, earliest, Chapter VII, 53; nomenclature,, Chapter VI, ... 39
No. I, Archaean, Chapter XVI, ... 165
No. II, Chazy and Trenton limestone, Chapters XXIV to XLIV, 298; fossils, Chapter XVIV, 118; hydraulic cement quarries, Chapter XXIX, 337; limonite ores, Chapter XXX, 341; limonites, Centre county, Chapter XXXII, 372; magnesian beds, Chapter XXVIII, 327; mechanical deposits, Chapter XLIII, 501; Nittany Valley, Centre county, Chapter XXXI, 365; oil and gas in, Chapter XLII, 492; thickness, Chapter XLI, 485; trap dykes, Chapter XXXVIII, 451; white limestone and marbles, Chapter XXXIX, 467; black marble, Chapter XL, 482; zinc, lead and barium, Chapter XXXVII, ... 436
No. III, Utica and Hudson River slates, Chapter XLV, 525; thickness, Chapter XLVI, 557; character of, Chapter XLVII, 562; mineralogical poverty, 570; neither oil nor gas in, 571; synclinal belts in, 286; roofing slate beds, Chapter XLVIII, 574; slate quarries, Chapter XLIX, 588; fossils, Chapter L, ... 617
No. IV, Oneida and Medina sandstone, Chapters LI to LIV, 625; conformity upon No. III, 706; thickness of, 627, thin southward, 649; topography in Middle Pennsylvania, Chapter LII, 681; names of mountains in, 681; ravine system in, 698; anticlinal vaults restored, 699; synclinal mountains in No. III, 278; difference between synclinal and anticlinal knobs in, 692; crests, single and double, 695; difference in height of mountains, 696; Keel mountains, 697; Medina, model of upper surface, 703; worthlessness of, Chapter LIII, 711; lead mines of, 678; in New Jersey, 676; in New York, 677; at gap above Harrisburg, 637; at Schuylkill water gap, 643, 673; at Susquehanna gap, 643, 669; at Juniata

gaps, 642; Bald Eagle gaps, 649; Wills mountain gap, 649; at Orbisonia, 653, at Spruce Creek gap, 655; at Tyrone gap, 657; at Mill Hall gap, 659; at Williamsburg gap, 661; in Bedford gaps, 661; in Clinton, Centre and Lycoming counties, 667; along the Great Valley, 669; in Delaware Water gap, 675; in Lehigh Water gap, ... 674

No. V, Clinton, Niagara, Salina, Chapters LV to LXIII, 721; no fossils in red rocks, 731; exceptions, 761; red color of, 725; thickness of, 728; limits of, 726, 728; outcrop water channels of, 729; at Delaware Water gap, 730; at Lehigh Water gap, 731; at Port Jervis, 733; at Schuylkill Water gap, 733; at Susquehanna Water gap, 735; on North Branch of Susquehanna, 727; in Montour Ridge, 737; along Big Fishing creek, 741; on Lower Juniata, Chapter LVII, 754; section east of Jersey Shore, 865; section at Muncy, 866; in Union, Snyder and Northumberland counties, Chapter LXIII, 870; in Centre, Clinton and Lycoming counties, Chapter LXII, 858; in Bedford and Fulton counties, Chapter LX, 838; in Blair county, Chapter LXI, 849; fossils of the formation, Chapter LXIV, ... 877

No. VI, Lower Helderberg, Chapters LXV to LXXIV, see Lower Helderberg, 898; limit of Silurian and Devonian, ...899 to 904

No. VII, Oriskany, Chapters LXXV to LXXXI, see Oriskany, 1034; on lower Juniata and in Fulton county, Chapter LXXVIII, 1084; supposed division between Silurian and Devonian, 1039; Caudagalli and Schoharie Grits, 1040; Pennsylvania outcrop of, ...1038

No. VIII, Chapters LXXXII to CV, ...1143

No. VIIIa, Upper Helderberg, Onondaga and Corniferous limestones, Chapters LXXXII and LXXXIII, see Corniferous, 1143; on Selinsgrove anticlinal, 1178; on North branch of Susquehanna river, 1180; lime shales and limestones of, 1185; Corniferous on Middle Juniata, 1188; in Bedford and Fulton counties, ...1189

No. VIIIb, Marcellus shale, Chapters LXXXIV to LXXXVI, see Marcellus, ...1194

No. VIIIc, Hamilton, Chapters LXXXVII to LXXXIX, see Hamilton, ...1236

No. VIIId, Tully limestone in New York, Chapter XC, see Tully, ...1307

No. VIIIe, Genesee black shale, Chapter XCI, see Genesee, ...1323

No. VIIIf, Portage, Chapters XCI, XCII, see Portage, ..1336

No. VIIIf, g, Girard shale, Erie county, Chapter XCIII. 1367
No. VIIIg, Chemung, Chapters XCIII to CV, see Chemung, ...1367 to 1545
No. IX, Chapters CVI to CX, see Catskill, ...1567
No. X, Pocono, Chapters CXI to CXIII, see Pocono, ...1629
No. X-XI, Mountain limestone, Chapter CXIV, ...1789
No. XI, Mauch Chunk red shale, Chapter CXV, see Mauch Chunk, ...1805
No. XII, see Pottsville conglomerate, Chapters CXVI, CXVII, 1853; coal measures, 1920, 1949, 1951; in Bernice Coal basin, 2009; in Eastern Middle Anthracite field, 2023, 2025, 2029, 2030, 2032, 2035, 2038, 2042, 2044; in Northern anthracite field, 1949, 1953, 1959, 1966, 1975, 1984, 2000; in Southern anthracite field, 2075, 2078, 2082, 2087, 2095, 2106, 2118, 2127, 2133, 2133, 2139; in Western Middle anthracite field, ...2050,2054,2059,2063,2068
No. XIII, Chapters CXVIII to CXXV, see anthracite and carboniferous, ...1916
No. XIV, Barren measures, Chapters CXXVI, see carboniferous, ...2406
No. XV, Upper Productive measures, Chapter CXXVIII, see carboniferous, ...2443
No. XVI, Upper Barren measures, Chapter CXXVIII, see carboniferous, ...2526
Fort Hill mine, Pittsburgh coal in, ...2491
" Plain, 1147; road, 1042; Cooperstown road quarries, ...1043
Forty Fort colliery, Northern coal field, ...1989,1990,1992
Fossils of No. IV, Chapter LIV, 714; of No. VI Lower Helderberg, Chapter LXXIII, 1005; of No, VII, Chapter LXXXI, 1125; of Hamilton, Chapter LXXXIX, 1283; of Chemung, Chapter XCIX, 1471, 1373, 1383; of No. X Pocono, 1675; Pocono (Tipton Run), 1689; Waverly rocks, 1779; of the coal measures, plates 537 to 595, after page 2588; of Genesee No. VIIIe, 1322; of Three Springs and Orbisonia, 1101; of Caudagalli Grit, 1139; of Schoharie Grit, 1141; of Corniferous, 1192; of Marcellus in Western New York, 1198; of Tulley limestone, 1312; of Portage, ...1343
Fossil ores in Blair county, 853 to 857; block ore in Bedford county, 848; in Frankstown bed, 847; in Bedford county, 841 to 848; outcrop brown hematite, 845 to 847; sand vein bed, 808; sand rock, 810; of Clinton, 749 to 752; Montour Ridge, 750; earliest fish, 756, 764; in Danville bed group, 813; Bird's Eye bed, 816; Saltillo bed, 824; in Upper Hamilton, ...1267
Foster's bank, Darlington coal C" at, ...2389
Four Foot coal bed, in Northern coal field, ...1964,2006
"   "   "   " in Southern field, ...2112,2122,2131
"   "   "   " in Western Middle field, ...2057

INDEX FINAL SUMMARY REPORT. 37

Fourteen Foot coal bed, Northern coal field, Pittson bed, ...1980
Fourth Bituminous basin, in Elk county, ...2283
" coal bed, Northern coal field, Clark bed, ...1979
" (Dudley basin), Broad Top coal field, ...2175
" or Peter's mountain, ...1651
Fouse's Pincher tunnel section, No. V, ... 830
Fox mine, ...2541-2545
Fractures and faults, ... 859
Franklin colliery, ...1987,1989,1992 1994,1995,1996,1997
" county, limonite mines and banks, 205, 231, 238, 248, 341;
  Path Valley mines, 357, 361, 409; Pond limonite banks of,
  351; limestone quarries, 311, 326; section across, 296; mines,
  205; zinc mine, New Jersey, ... 469
" quarry, roofing slate of No. III, 606; (type of rocks), Ar-
  chaen, 72; (wells at), 26; Conway well below, ... 27
Frankland's analysis of sea water, ...20,21
Frankstown anticlinal, Oriskany S. S. on, ...1119
" bed, No. V iron ore, description of, ... 847
" Cambria Iron Company's slope near, ... 895
" Cove, No. VIII rocks in, ...849,850,1355
" mines, 850, 854, 859; slope mine, ... 855
Frazer's P. analysis of Dysart mine iron ore,......................... 854
" P. Lancaster county survey, ... 555
" Rep. CC. Rep. C., ...473,474
" section and cross sections, 129, 133, 144; maps, ...127,134.167,171
" sections of South mountains of Cumberland and Franklin
  counties, ... 45
Frazier, J. opening, Pittsburgh coal, ...2540
Fredericktown axis, description of in Beaver county, ...2396
Freeport coal group, general description of in Clearfield county,
  2212, 2213; limestones, 2158, 2161, 2293, 2365, 2367, 2371, 2398;
  lower coal D, 2235, 2236, 2246, 2249, 2255, 2260, 2288, 2293, 2297,
  2299, 2318, 2350, 2366, 2371, 2375, 2385, 2386, 2394, 2398; D, in the
  Blairsville basin, 2331; lower coal bed D, in Ligonier basin,
  2322; in Marion-Saltsburg basin, 2340; in Centre county,
  2194; sandstone, 2294, 2318, 2382, 2398; Freeport upper coal
  bed E, 2222, 2223, 2233, 2246, 2249, 2256, 2259, 2293, 2294, 2295,
  2296, 2318, 2320, 2363, 2346, 2370, 2374, 2381, 2385, 2392, 2394, 2397;
  upper coal E, in Blairsville basin, 2329; in Marion and
  Saltsburg basins, 2337; in Centre county, 2194; in Morris-
  dale-Philipsburg region, 2207; upper fire clay at Salina, 2346;
  at Bolivar, ...2365
French creek (Aboriginal relics), ...1509
" " 262; magnetic mines, 267; copper mine, ... 267
" " 1506, 1507, 1522; North Branch and branches, ...1500,1512
" Portland cement, ... 339

38　　　GEOLOGICAL SURVEY OF PENNSYLVANIA.

Frey's pit, coal E, Indiana county, ...2338
Friedensburg, 236, 237, 438; zinc mines, ... 344
Friends' cove ore and limestone, Chapter XXXV, ... 419
Frill's quarry, No. II limestone, ... 311
Fritz's, Horst's and, quarry, limestone No. II, Lebanon Valley, ... 315
Fritz, Peter, quarries, York county marble, ... 473
Fritz's Island, 355, 464; mines, ...103,269
Frost Station section, Pittsburgh ore group, ...2435
Fry bank, bed D, Indiana county, ...2339
Fry's (Peter) farm (quarry), slate of III, ...589
Fuller bank, bed D, Indiana county, ...2324
" mine section, Mercer group in Jefferson county, ...1887
Funk bank, bed E, section at, Cambria county, ...2238
Funk's quarry, Bossardville limestone beds at, ...956
Fulmer (Henry) quarries, roofing slate of III, ... 585
Fulton county, Catskill in, 1613; colored map in T2, 1574; limestone coves, 425; No. IV measurements of, 631; and Bedford counties, No. V in, Chapter LX, 838; county, No. VI in, Chapter LXXI, 983, 996 to 998; No. VII in, 1084, 1090, 1091, 1092; No. VIIIa in, 1189; No. VIIIc in, 1281; No. XII in, 1864; Portage in, 1356; Rep. 1405, 1407, 1413, 1425, 1431; Rep. T2, 424; section, Catskill and Chemung rocks, 1412; and Pinkerton mines, Pittsburgh coal at, ... 2514
Furnace quarries at West Conshohocken, old gneiss of Buck Ridge,... 81

G.

Gabbro character and occurences, ...2625
Gable mine, Upper Freeport, coal E, Indiana county, ...2330
Gaines basin, ...1871-2,2189,2264-75-76
Gamble opening, bed C' at, in Indiana county, ...2326
Gamma coal bed, Eastern Middle coal field, ...2029-33-6-7-9-43-5-6
Gap hills, Gneiss belt, ... 78
" mine range, ... 453
" nickel mine, Lancaster county, ... 288
" of Delaware river above Easton, ...82,105
" of Bushkill, ... 105
Garber's quarry, fossils of No. VI in, ...1027
Gardner's mine, bed D in, ...2323
Garfield mine, Pittsburgh coal, ...2552
Garland, Warren county, ...1508-28-29
" section (Cotter farm), ...1528
" conglomerate, ...1895
" spirifer bed, ...1773
Garlow, John opening, section of Pittsburgh coal, ...2510
Garman's quarry, No. II limestone of Dauphin county, ... 320

INDEX FINAL SUMMARY REPORT. 39

Garrison's cut, Connoquenessing Upper sandstone, ...1888
" opening, Washington coal bed at, ...2575
Gaskins' opening, bed D, Indiana county, ...2342
Gaspe, ...160,1137-9
Gate coal bed, Pottsville division, Southern coal field, ...2115
" **Ridge anticlinal,** 2074-95
Gatesburg Ridge anticlinal, Centre county, ... 367
" bank, limonite ore, Centre county, ... 379
Gatiss mine, bed B section, ...2265
Gaut pit, Pittsburgh coal, ...2514
Gaylord colliery, Northern coal field, ... 1995
Gehret's quarry, No. II limestone, Lebanon county, ... 311
General sections, definitions, Chapter IV, ... 30
Genesee black slate, No. VIIIc, Chapter XCI, ...1323
" river gorge, ...1243
" fossils, ...1332
Geological dimensions, Chapter III, ... 22
" knowledge, Chapter I, ... 1
" map of 1841, ... 870
" **section across Vermont and New Hampshire,** 58
" time, Chapter II, ... 5
George coal bed, Diamond or "I" coal; Northern coal field, ...2307
" mine, Pittsburgh coal, Armstrong county, ...2459
Georgetown sections, ...964,1256
" anticlinal, ...1256
" quarries near, ... 963
German bank, Bloomfield ore clays at; limonite, ...414-6
Germania Oil Company, ...1471 to 79
Germantown, limestone quarries near, ... 470
" McKinney's quarry, ... 114
Gibson well, Oil Creek Lake Group No. X, ...1767
Gingrich's quarry, No. II limestone, Dauphin county, ...319,320
Girard quarry, No. III roofing slate, ... 606
" colliery, Mammoth bed, Western Middle coal field, ...2061
" shale, ...1367
Gilt's flagstone quarry, No. III slate, ... 611
Glacial drift in Northern coal field, ...2016
Glade run axis, ...2348-53
Glanden's run coal field, Allegheny river coal series, ...2242
Glass sand quarries, Mapleton range, ...1103
Glen Lyon colliery, Northern coal field, ...2003 to 2006
" Mayo coal bed, in No. XII rocks, ...1884
" Mayo colliery, described Wilmarth tract, ...1884
" **White mine, bed C',** 2230
Glendon Iron Company's mine, limonite ore, ... 232
" " " works, ... 802
Glendower colliery, Southern coal field, ...2085

Glen's Falls marble quarry, ... 482 to 484
Glenville section, Pocono s.s. Crawford county, ... 1755.
Glenwood Coal Company; bed D, ... 2324
 "   colliery, Northern coal field, ... 1963-65
Gneiss, newer Philadelphia belt, Chapter XI, ... 118
Godfrey's quarry, Portage flags, ... 1354
Goheen bank, bed E, ... 2374
Gold, traces of, near Yardleyville, in New Red, ... 2633
  "   mine gap section, Southern coal field, ... 2143
Goodman's quarry, Third oil sand, ... 1516
Good Spring colliery, Southern coal field, ... 2127-9,30-1-2
Gordon colliery, Western Middle coal field. 2060
Gorge of Neshiminy and Pennypack, ... 79
Gorman, Alex. bank, bed D, ... 2324
 "   James, opening, bed C, ... 2325
 "   tunnel, Southern coal field, ... 2097 to 2101
Goshen, trap dyke in No. II near, ... 453
 "   tunnel, ... 817
Goul's quarry, No. II limestone, Berks county, ... 310-13
Gowen colliery, Eastern Middle coal field, ... 2036
Grafton quarry, No. VI limestone, Powell Iron Company, ... 993
Graft's mine, coal E, Indiana county, ... 2338
Graham bank, coal D, Indiana county, ... 2324
 "   mine, Pittsburgh coal, ... 2463
Grand Canon fish, ... 769
 "   Central quarry, roofing slate No. III, ... 586
 "   river, Canada, ... 889
Grant mine, Pittsburgh coal section, ... 2541-2
Granville gap section, shade Mt. Danville fossil ore, ... 812
Grassy Island colliery, Northern coal field, ... 1957-61-4-5
Grable's mine, ... 2535
Graver and Co. M. mine, Pittsburgh coal, ... 2530·
Great Aughwick Valley, 779-81, 1222; Belt level, 1542; Bend of Susquehanna, 1416, 1417-37-42-3-8, 1581-7; break in West Virginia, 1541; Britain, order of formations, 771; Horseshoe dyke, 458 to 463; Ironstown Ridge, 459; limestone, 2451-74, 2501-16; Manitoulin island, 881; North mountain, 1570-1, 1607-51; Savage mountain,.. 681
 "   Valley, geology of, 270, 500, 553, Chapter XXII. 270; two belts, slate and limestone. 274; present site, 566; section, 577; mineral wealth. 570-71; limestone formation, No. II, 70, 116, 298, Chapter XXV, 309; quarries, 299, Chapter XXVI, 309, Chapter XXVII. 324; limonites, Chapter XXX, 341; slate belt, 562-5, 669; ore horizons, 365; sink holes, 425; limestone coves in slates, 283; trap dyke, 455; volcanic action. 564; great mines, 361; why there is no coal in it. 294; mineral worthlessness of the North mountain in. .... 712·

Greencastle, 273-87; level in, ... 274
Greene county group, ...43,47,2566
"   "   " in Greene and Washington counties, ...2530
"   "   No. XIV in, ...2446
"   "   No. XV in, ...2524
Greendale axis, ...2348
Greenfield postoffice trial well, Chemung rocks. 1500
Green mountain coal basins, 2044, slope, Eastern Middle coal field,...2045
"   mountains of Vermont, ...45,61,159-60,119,552
"   Pond mountain, ...1365
"   Ridge colliery, Northern coal field, ...969
"   Springs mine, Pittsburgh coal, ...2545,2546
Greensburg basin, description of, ...2435,2565
"   "   No. XV in, ...2497
"   Coal Company's mine, Pittsburgh coal, ...2499
Greenwalt coal bed, Southern coal field, ...2132
Greenwood colliery, Northern coal field, 1970-7; tunnel, Southern coal
    field, ...2058
"   furnace, fossil ore section, ...833-5
"   ore banks. 421
"   fault, ...421
"   section in No. X, ...144
Gregory's run, ...823,1109
"   " section, No. V rocks, ...824
Gresh mine, impure coal in No. XI in Cameron-Elk. 1839
Grey Coal bed, Southern coal field. 2144-5
Griffith mine, bed C', 2326-31; opening, ...2334
"   Jones slope, Southern coal field, ...2113
"   Uplinger &, quarry, roofing slate No. III. 601
Griswold gap conglomerate, ...1643-7-9
Groninger's quarry, limestone No. II. 315-6
Grosser bore hole, bed D, ...2227
Grossier beds of Flanders, ...328
Grove quarries, ...303,422,589,750,947-53,1009-24,1133
"   bank, bed C', Somerset county, ...2246
"   tunnel, ...939-45-53-59,1025
"   " section, No. VII, Montour Ridge, ...1074
"   (Patterson) district; No. V fossil ore in, ...825
Grubb-Kurtz district, No. V fossil ore in, ...828
Grube's quarry, No. VI Bossardville limestone, ...930
Gruber and Bowman pits and kilns at Palmyra, No. II limestone, ...318
Gudlin quarry, limestone of No. II. 311
Gulf Brook, ...1444
"   "   (Leroy) section. ....1448 to 1466
"   creek, ...1452
"   "   section, ...1469
    3*

42 GEOLOGICAL SURVEY OF PENNSYLVANIA.

Gulf of Mexico, ... 292
" " line of mountains from the St. Lawrence to, ... 48
Gwynedd shales, 2610; fossils, 2611; trap, 2619-23; coal, 2629, soil, ...2626
Gwynn's, Jones &, falls, typical biotite gneiss at, ... 130
Gypsum, beds of No. VI in New York, ... 913

H.

Hacklebarney tunnel, bed B in Southern coal field, ... 2088
Hagerman's run gap, ... 865
Hagerstown, 273; level in, ... 274
Half Falls mountain, 1087, 1187; gap, 1185-8; range, 1273; section, ...1187
Hall, C. E., report on Adirondack rocks. 109
" " list of Hamilton fossils,, ...1303
" " trap near Trenton Cut-Off, ... 2620
" James, sections, ...1036,1159,1242,1308-37
" " MS. list of type specimens, ... 1202
" sketch of a Waterloo quarry, ... 1163
" mine, Pittsburgh coal at, ... 2521
" quarries; synclinal fold at, in No. VI limestone, ... 981
Halstead colliery, Northern coal field, ...1875-8
Hamilton formation, No. VIIIc, Chapters LXXXVII to LXXXIX.
 1237 to 1283; in Pennsylvania, Chapter LXXXVIII, 1170;
 in New York, 1257; its changeful character, 1244; on main
 Susquehanna, 1255; north of Montour's ridge, 1264; in
 Perry county, 1266; in Juniata county, 1275; in Snyder,
 Mifflin and Huntingdon, 1276; in Bedford and Fulton, 1281;
 lower sandstone, 1280; lower shale, 1280; upper shales, 1277;
 middle shales, 1279; upper sandstone, 1278; upper fossil ore,
 126; fossils, Chapter LXXXIX, 1283; quarries, 990;
 mine, ...2297,2305
Hamlin mine, Alton lower coal at, ... 1877
Hampton mine, Pittsburgh coal, ... 2528
Hanover Hogback anticlinal, ... 2002
" iron works, ... 423
" tunnel, Northern coal field, Red ash coal, ... 2003
Harbor Mountain; Pocono s.s. in, ... 681
Harden, E. B. models of the second survey, ...703,1917,2000
" " surveys by, ...387,2008
Harleigh colliery, Wharton coal, Eastern Middle coal field, ... 2033
Harper's colliery, Southern coal field, ...2112-13
" Ferry, ...145-6,271
" " level of Potomac at, ... 272
" quarry (old), No. III slate, ... 600
Harriet Lane mine, Kelly bed section, Broad Top coal field, ... 2170

Harrisburg, gap above, 637-69; quarries opposite, 497; well bored
    north of, 571; levels in, 272-4; section, 296; furnaces...242,258
Harrison well, Potter county, Chemung rocks, ...1478
Harrisonville axis, ...2380
Harrity open cut, Morrison Cove limonite, ... 416
Harry, E. colliery, Northern coal field, ...1988 to 1992
Hartley's Mill bank, Sewickley coal, ...2562
Hartlieb's quarry, No. II limestone, Lebanon county, ... 314
Hartman's cove, Monroe county, ...429,916
Hartman No. 2 mine, Pittsburgh coal, ...2471
Hartzlog valley, formation No. V in the, ...819-22
Haskill well, lower Chemung shales in, ...1485-7,1542
Hastings colliery, bed E, ...2241
    "   mine, bed A, 2314; Pittsburgh coal, ...2543
    "   district, Cambria county; described; bed E. 2238-40
Hathaway's quarry, Tully limestone in, ...1307
Haun's bridge, ...1409-10,1555,1611
    "   " section, Chemung up. congl., ...1551,1611-12
Haviland's gold, New Red of Bucks county, ...2633
Hawbecker's quarry (Williamson quarry) No. II limestone, ... 326
Hawk, P. mine (Adrian mine), bed D in, ...2302
Hawthey mines, bed B, ...2234
Haycock mountain trap, New Red, ...2618-22-23
Hay's mine, bed C', Somerset county, 2247; Pittsburgh coal, ...2546
Hazardville, Oriskany sand stone near; paint ore belt, ...1061-3-9
    "   sections, Oriskany sand stone, ...1055-65
    "   paint tunnel; hydraulic cement near, ...1175
Hazel Dell colliery, Western Middle coal field, ...2063-5
Hazen and Burton mine, Pittsburgh coal, ...2522
Hazle Brook colliery, Eastern Middle coal field, ...2039-40
Hazleton coal basin, ...2038
Hazlett mine, bed E, ...2338
Hebron mountain, geology of, ...1464-77
    "   anticlinal, ...1463-4-78
    "   well; Catskill rocks in, Potter county, ...1479
Hecla furnace cross section, ... 370
    "   mines, 1 and 2, Connellsville basin; Pittsburgh coal, ...2488
Heckscherville valley (or Mine Hill) basin, ...2081
Heidelberg township, Lehigh county, quarries in No. II. 608
Heilpin, Angelo, fossils in Wyoming Valley limestone, ...2021
Heimbach mine, bed C', Somerset county, ...2259
    "   quarry, No. III roofing slate, ...586,602
Hein and Glasmire slope, Southern coal field. 2104
Heinmeyer bank, bed C', ...2247
Heishley's ore bank, Hamilton rocks, Perry county, ...1268
Heister coal bed, Southern coal field, ...2144-5

Helderberg hills, ...904-5,1035
Helfenstein colliery, Western Middle coal field, ...2060
Hellam quarries, Scolithus linearis fossils in, ... 191
Heller's quarries, Bossordville limestone No. VI, ... 932
Hellman's pit, bed B, Indiana county, ...2336
Helman's quarry, No. III slate, ... 598
Helvetia mines, bed D in Jefferson county. 2298,2304
Hematite ores in Primal Upper slate, Chapter XX, ... 205
Henderson's, Dr. cross section, ore horizons, Centre county,........... 386
    " fault, Oriskany sand stone in, Fulton county, ...1038
    " quarry, marble in No. 11, ... 473
    " mine, Washington coal bed, ...2576
    " survey of 1839 to 1840 and 1849, ...1091,1187
Hendricks' drift, Northern coal field, ...1956
Henrietta mines, Limonite of Blair county, ...341,361,413-4
    " (Leather Cracker) anticlinal, ... 409
    " mine and shaft, bed B; Cambria county, ...2227-8
Henry Clay colliery, Western Middle coal field, ...2071
Henry colliery, Northern coal field, ...1990-95-96
Henry's quarry, No. III slate, 595, 600; Olean Congl., ...1901
    " valley, Perry county, ... 682
Hering, Rudolph, topographical map, Philadelphia district. 2597
Herminie shaft, Pittsburgh coal. Westmoreland county, ...2513
Heron's drift, Diamond bed, Southern coal field, ...2113
Herr's sink or quarry, No. II limestone, Lebanon county, ...315-6
Herschell's estimate, water in the world ocean, ... 20
Hess' quarry, Bossardville limestone beds in, ... 952
Hetherington mine, Waynesburg coal bed, ...2556
Hewston's opening, Waynesburg coal bed. 2561
Hickerman ledge, No. IX Catskill; New Milford sand stone, ...1581
Hickman pit, Pittsburgh coal, ...2563
Hickory Coal Company, Sharon coal mined by the, ...1908
    " slopes, Southern coal field, ...2110
    " ridge, ...394-5,367
    " " colliery, Western Middle coal field, ...2069-70
    " swamp colliery, Western Middle coal field, ...2069-70
Hicks' cement quarry, in No. II limestone. Centre country, ... 340
    " ore bank, Marcellus ore, Huntingdon county, ...1228-9-31
Highberger, Ezra, pit. Pittsburgh coal, ...2513
Highland Coal Company, Eastern Middle coal field, ...2027
    " colliery, Eastern Middle coal field. 2029-30
Highlands of East Pennsylvania, 159; of Berks, Lehigh and North-
    ampton. 78, 82-3, 144-6, 166, 236; of South mountain, or Blue
    Ridge, 106-7, 291; of South mountain or Blue Ridge, gap
    in. 70; west of Delaware river, 112; of Pennsylvania, 61
    to 63, 70, 76, 101-8-59; belts. 116

INDEX FINAL SUMMARY REPORT. 45

Highland section, olenellus zone in the, ... 192
" township, Elk county oil fields, ...1485
Hill farm opening, Pittsburgh coal, Connellsville basin, ...2492
" Frank A., ...1069,1938-99,2013 17 53 58-62
" J. mine, Waynesburg coal, ...2557
" of Chickies sandstone, New Red district, ...2602
Hillman coal bed, 1982-95; colliery, 1996; vein colliery, 1996; slope or G
   coal bed, Northern coal field, ...2006
Hillsdale colliery, 1980; mine, Pittsburgh coal, ...2550
Hillside Coal and Iron Company's borings, Northern coal field, ...1955
Hindale mine, Pittsburgh coal, ...2552
Hitchcock section across Vermont and New Hampshire, ...58,159
Hitner's quarry, black marble of No. II, ...482
Hoerner's quarry, No.VI limestone, Clinton county, ...1004
Hoffman bank, bed D, Indiana county, ...2324
" pit, ...2524
" S. B., house, quarry near, No. III slate, ... 600
Hogback ridge, oriskany sandstone, ...1117
" Walpack Bend cross section, Oriskany sandstone, ...1049
Hokendauqua quarry, cleavage planes of No. II, ... 306
Hoke's quarry, limestone No. II, ... 319
Holden colliery, Northern coal field, ...1976
Hole mountain, ...278-9,710
Hollenback colliery, ...1987-92-5-6-7
Holler's quarry, No. II limestone. 321
Hollidaysburg anticlinal, ...1119
" quarries, No. VI limestone, Blair county, ...1001
" Cambria Iron Company section, Clinton No. V, ... 852
" Frankstown district, block ore bed unworked, ... 857
Holmes coal bed in Western Middle field, ...2057-61-66-71
"   "   "   " Southern field, ...2100-12-22-31-37
"   "   "   " Heckscherville basin, ...2085
Holsopple mine, bed C', Cambria county, ...2235
Home shaft No. 2, Sharon coal in, ...1909
Homer-Ulysses, Catskill anticlinal valley, ...1435
Homewood sandstone, No. XII, ...1859-63-8-86-96-7,1910-12-15,2155,2315-19
Hommer mine, bed B, Cambria county, ...2231
Honesdale cliff, Montrose red shale, No. IX, ...1581
" No. IX, Catskill rocks, ...1585
Honeycomb bed, Umbral ore group, No. XI, ...1848
" " analysis, Umbral ore group, No. XI, ...1852
Hopewell-Yellow creek section, Pocono sandstone, Catskill, ...1615
Hope works fire clay analysis, Clearfield county, ...2216
Horner mine, Pittsburgh coal, ...2534
" & Roberts mine, Pittsburgh coal, ...2537
Horseshoe dykes, trap in limestone No. II, ...451-9-60-2-3

GEOLOGICAL SURVEY OF PENNSYLVANIA.

Horse valley limestone and ore, Chapter XXXV, ... 419
Horshamville trap rubbish, ... 2620
Hosmer run, 1508; level, 1542; wells, 1491; aboriginal oil pits, ... 1509
Houck bank, Cook seam, Broad Top coal field, ... 2182
Houser's drift, Southern coal field, ... 2121
Houtzdale-Philipsburg basin, Centre county, described, ... 2209
Howard-Jacksonville cross section, sand-lime beds of No. II, ... 370
Howard's, 1122; quarry, 1498, 1516-18; section, Venango oil group,....1515
Hubler bank, section, analysis, bed B, Clearfield county, ... 2216
Huffert's, Oriskany section, along Delaware river, ... 1057
Huffman, J. J., mine, Waynesburg coal, ... 2556
" " analysis, Freeport lower coal D, ... 2300
Hugus & Job pit, Pittsburgh coal, ... 2509
Hukill well, Chemung VIIIg rocks, ... 1487
Huling's oil well, Chemung VIIIg rocks, ... 1482-7
Humboldt colliery, Eastern Middle coal field, ... 2039
Hunt colliery, Northern coal field, ... 1990-92
Hunter opening, Pittsburgh coal, ... 2488
Huntingdon county, **g**eology of, 500, 1099; sections, 833, 983, 1619; sections Coffee run, 1610; Spruce creek anticlinal, 367; Hickory ridge anticlinal, 367; Broad Top, 489; No. IV measurements of, 631, 645; No. VI 983, 991 to 996, Chapter LXXI, 983; Marcellus ore, 1228; No. VII, 1093-7, 1100-03; Portage, 1359; No. VIIIc, ... 1276
" and Bedford district, ... 1833
" " " No. XII, ... 1862
" valley, No. V, Chapter LIX, ... 819
" " description, ... 820
Hurd coal bed, Gaines basin, Tioga county. 2275
Huronian-system, described, Chapter XV, ... 152
Hutchinson coal, Northern coal field, ... 1997
Hyde Park collieries, Northern coal field, ... 1970
Hyndman, ...998,1029,1116-17
" section, Wills creek, ... 1615
" " No. VI, ... 997
" " No. VII, ... 1110-17
Hyner, section, Chemung VIIIg to No. XIII rocks, ... 1620
" axis, ... 2190
" run oil well record, ... 1621

I.

Identification of layers, New Red rocks, ... 2605
" of rock beds, New Red rocks, ... 2603
Ihler, John, opening, bed D, ... 2209
Illinois, little, section, No. V, Clinton Upper shales, ... 757-8

Illinois and Indiana coal basin, mountain limestone in, ... 1791
India, glacial traces, ... 51
" madstones or snakestones, ... 566
Indian level mines, ... 678
" painted figures, ... 1442
" Standing Stone, ... 1107
Indiana anticlinal, ... 2162,2302
" axis, ... 1888,2316-36
" oil and gas districts, ... 436
" county, No. XII in, ... 1889
" " No. XIII in, ... 2315
" " No. XIV in, ... 2423
" " No. XV in, ... 2465
Industrial slate quarry, roofing slate of No. III, ... 606
Ingleside mine, bed B, Somerset county, ... 2250
Inoculate ridge, Oriskany sandstone, important in, ... 1088
Iron ore, Marcellus, Chapter LXXXVI, ... 1220
" Company's railroad cuts, ... 745
" Hill, ... 1088
" sandstone of Clinton, No. V, ... 749,755,803,815
Ironstone Ridge dyke or Cumberland Valley dyke. 458-9
Ironton district, described, ... 345-63
Irvin, Alex., bank, bed D, Indiana county, ... 2325
" farm well record, Venango oil group, ... 1539
Irwin, gas coal basin, described, ... 2501
Isle, La Motte, ... 483-4
Ithaca wells, Genesee rocks in, ... 489,1332
Ivile mine, Pittsburgh coal, ... 2552

J.

Jacks mountain range, ... 872
" " section, ... 669
" " outcrop, ... 659
" " anticlinal, ... 706
" " map of south flank, ... 699,795
" Narrows, ... 655
" " section, ... 645,799
Jackson, fault, ... 402-3
" quarry. Slate No. III, 594; building stone in, ... 1753
" mine, Pittsburgh coal at, ... 2190
Jacksonville valley group, ... 378
" -Howard cross section, ... 370
Jacob's slope, Pittsburgh coal, ... 2523-60
Jeddo collieries, Eastern Middle coal field, ... 2033
" drainage tunnel, Eastern Middle coal field, ... 2034

Jefferson Coal Company, Mercer Upper coal, ... 1887
"   county, ... 1840
"   "   No. XII, ... 1886
"   "   No. XIII, ... 2292
"   "   No. XIV, ... 2422
   mine, Mercer coal, 1887; Waynesburg coal, ... 2561
Jenny Jump range, rose colored marble beds in, ... 109,118
Jericho Mountain trap, New Red district, ... 2619 to 2626
Jermyn collieries, Northern coal field, ... 1959-63 to 65-77
"   Priceville division, Northern coal field, ... 1958
Jersey Shore gap and sections, Nos. IV and V, ... 667,865
Jesse Bell mine, Waynesburg coal, ... 2561
"   Cook bank, Cook bed, Broad Top coal field, ... 2179
Jock coal bed, Southern coal field, ... 2092
Johnsonburg-Elk county coal basin, ... 1837
Johnson Run basin, 1884; sandstone, ... 1882-4,2282
"   " and Smyrna sandstone, ... 1862
"   " synclinal, ... 2289
Johnstown (or Second) basin, 2253-6; No. XIV in, 2421; district, model
   of, 703; analyses, 856; ore bed, 2369; slope mine, 433;
   cement bed, 2160-96, 2307-18-28-67-76; gap, 28; Cambria Iron
   Company's works, ... 412-17
John's pit, Sewickley coal, ... 2564
" & Rich slate quarries, No. III, ... 585
"   run gap, ... 143
Jollytown coal, ... 2568-72
"   sandstone, ... 2573
Jones opening, Pittsburgh coal, ... 2509
"   mine, Pittsburgh coal, ... 2548-55
"   & Company's quarry, Peach Bottom slate, ... 138-9
"   & Gwynn's falls quarries, biotite gneiss, ... 130
"   & Laughlin well, Venango oil group, ... 1543
Jory quarry, No. III slate, ... 593
Jugular coal basin, Southern coal field, ... 2082
Julien's 'Genesis,' 155
Jumbo mine, Pittsburgh coal, ... 2541
Junction quarry, No. III roofing slate, ... 606
Juniata county fossils, 1007, 1133; fossils, crustacean, 893; No. VI in,
   Chapter LXXI, 983, 989 to 91; No. VII in, 1071-84-90;
   Marcellus ore, 1221; No. VIIIc, ... 1275
"   region, ... 1549-72
"   river section, 969, 691; gaps, 401, 643, 661, 820-43-8; quarries on,
   1034; mines on, 237; upper, 1037-43, 1090, 1196; upper No.
   VII on, 1103; middle No. V on, Chapter LVIII, 779;
   middle No. VIIa on, 1188; central valley 1276; lower No.

INDEX FINAL SUMMARY REPORT. 49

  V on, Chapter LVII, 753; lower No. VI on, Chapter
  LXX, 967; lower No. VII on, 1084; lower No. IX Cats-
  kill, 1601; mouth, 296; Horse Shoe Bend, 688; exposures
  on, 40; fossil ore district, ... 789
" Narrows, Clinton No. V rocks in the, ... 812
Jurassic age, ... 18
Jura valley, formations of, ... 701

K

Kalmia colliery, ...2139,2143
Kane, Olean conglomerate at, ...1875
Karthaus-Renovo basin, ...2193
Kaskaskia limestone, thickness of. 1791
Kaska William colliery, ...2097,2099,2101,2102
Katy-Did colliery, ...1950
Kaufman's bank, Mercer lower coal at, ...1904
Kearney mine, Kelly coal in, Broad Top basin, ...2171
Kear's slope, Southern coal field, ...2112
Keel mountains of IV, topographical features described, ...697,698
" ridge, Mercer lower limestone in, ...1904
Kehoe bank, Pittsburgh coal, ...2486
Keim and Livengood mine, Sewickley limestone at, ...2464
Kelly seam, Broad Top basin, ...2166,2170
Kemble drift, Broad Top basin, ...2119
" slope, Kelly coal in, Broad Top basin, ...2170
Kemmerer, J. P. opening, Pittsburgh coal at, ...2508
Kendall mine, Sewickley coal bed at, ...2523
Kennedy opening, bed D, Jefferson county, ...2302
" S. C. opening, bed E, Indiana county, ...2337
Kennedy's granite quarry, Delaware county, ... 97
Kentucky coal field, ...1631
" colliery, ...2096,2098
" oil and gas districts of, ... 499
Kenyon College, Knox county, Ohio (well near), ... 1486
Keokuk group in Kentucky, ...1791
Kepple, S. V. bank, Pittsburgh bed, Greensburg basin, ...2500
Kerns quarry, No. VI limestone in, ... 981
Kersey Coal Company's lands, ...1883
Kesler openings, bed D, Jefferson county, ...2303
Kistenbaugh's tunnel, Palmer bed in, Southern coal field, ...2104
Kester's Meadow quarry, roofing slate of No. III at, ... 602
Kettle creek, or Oleona basin, ...276
" the, No. II limestone at, zinc mines at, ...427,444
Keyes section, ... 146
Key's mine, bed D, Jefferson county, ...2307

4

Keystone colliery, ...1963,1979,1980
" mine, Dagus bed B, at, 2284; Pittsburgh coal, ...2531
" (old) quarry, roofing slate of 111, ... 604
" tracts, Alton coal beds at, ...1882
" Zinc Company, 445; shafts, 445; land, ... 445
Khamzin (or desert wood), ... 468
Kidney coal bed, Northern coal field, ...1996
" ore analyses, ...1852
Kiester opening, Pittsburgh coal at, ...2490
Kimball's quarries, Lumbertown, grey S. S. of the New Red, ...2628
" 1591; run, 1591; quarries, ...1591
Kind, Brownscomb and, great quarries, Catskill rocks, ...1593
King mine, Dagus bed B, Elk county, ...2285
" mill sandstone of Perry county, 1560; fish beds, ...1561
Kinghorn bank, coal E, Indiana county, ...2338
Kigler drift, Southern coal field, ...2142
Kingston collieries, ...1988,1990,1993,1994,1995
" N. J. 583, 1137; cement beds, ... 887
" (quarries), ... 925
King's mill, 1563, 1560, 1603; sections near, 1558; fish beds and plant
beds, ...1561
Kimmel mine, bed C' at, ...2254
Kintz mine, Pittsburgh coal, analyses of, ...2487
Kinzua bank, bed D at, ...2339
" creek, fossils in No. VIII on, 1469; Chemung rock on, ...1432
Kipple, J. C. opening, Pittsburgh coal at, ...2510
Kirk and Baum slope, Southern coal field, ...2110
Kishicoquillis valley, 340,365,421,425,533,653,682,684,694,698,820,928,1039,1573,1823
" " limestone and ores Chapter XXXV, ... 419
" " anticlinal, same as Nescopec axis, ... 706
" " map, caverns and sinkholes in No. II, ... 425
Kister opening, bed E at, ...2252
Kister's, D. opening, Pittsburgh coal at, ...2508
Kittanning coal group, 2214, 2294, 2375, 2382; group in the Blairsville
basin, 2334; group in the Ligonier basin, 2325, 2367; group
in the Marion-Saltsburg basin, 2343; lower coal bed B,
2156, 2200, 2222, 2249, 2254, 2280, 2288, 2310, 2317, 2326, 2352,
2355, 2368, 2376, 2395; middle coal C. 2246, 2308, 2318, 2352,
2368, 2376, 2386, 2395; (New Brighton) fire clay, 2400; sandstone, description of, in S. Fayette county, 2368; upper
bed C' in Snow Shoe basin, 2197; upper coal C', 2222,2223,
2230, 2235, 2249, 2253, 2259, 2261, 2307, 2318, 2334, 2352, 2355,
2376, 2386, 2398; upper coal C' in Clearfield county, 2207;
upper coal C' in Centre county, ...2195
Kittatinny gap, Hamilton S. S. of No. VIII in, ...1255
" mountain, 633,637,647,676,695,710,713,731,917,935,1055,1060,1084,1263
1267,1211,1266

INDEX FINAL SUMMARY REPORT. 51

Klaproth, analysis of celestine by, ...1001
Kline's ridge quarries, Winfield, No. VI beds at, Snyder county, ... 985
Klopperdale valley, ore sandstone in, ...S11,813
Knepper bore hole, lower coal measures in, ...2227
Knight & Co. mines, bed D at, ...2240
Knob mine, Pittsburgh coal at, ...2554
Knobbly mountain, ...998,1114,1115,1117
Knorr's quarry, No. II limestone in, ...312
Knowledge, geological, Chapter I, ...1
Knox County, Ohio, P. Neff's gas well, ...1486
 " mine, Pittsburgh coal at, ...2472
Kober's copper in New Red of Bucks and Montgomery counties, ...2632
Koch, A. quarry, roofing slate of No. III in, ...599
Kohler's gap, in Thick mountain, ...2127,2129,2130,2131
 " (Mary) quarry, typical breccia limestone in Great valley, .. 305
Kreider's quarries, Annville (Lebanon valley) limestone No. II, ...317,318
Krock's quarry, encrinal stems abundant in, No. II, in Northampton
    county, ...308
Kuhn's opening, Pittsburgh coal in, ...2488
Kyle mine, Pittsburgh coal bed section at, ...2496

L

Lackawanna, red ash coal near, 1977; basin, disappearance of No. XI
    red shales in, 1827; basin, thickness of coal measures
    in, 1968; Coal Company, borings by, 1962; colliery,
    Northern coal field, 1965; Shickshinny mountain, 1651,1647
Lacoe, R. D. fossil collection from Campbell's ledge, Northern coal
    field, ...1950
Lafayette soapstone quarries, ...107,114,123,124
Lake district of Northwestern Pennnsylvania, ...1350
 " Erie district, 1377; region. 1116
 "   " Venango rocks towards, ...1506
Lambert opening, bed E in the, ...2247
Lancashire No. 3 and 4 mines, bed D at, ...2240
Lancaster county, Serpentine, 104; survey, Frazer, 555; Rep. C3,
    Frazer, 44, 78, 186, 219, 226, 227, 228, 229, 454; map,
    127, 136, 171, 203; thickness of No. II in, and else-
    where, Chapter XLI, ...485
 " section, Neffsville to Martinville, No. II limestone, ... 488
Lane ore bank (Oriskany), ...1097
Lance colliery, Northern coal field, ...1993,1995,1996,1997
Landisburg, fish bed S. S., 768; section of rocks at, ...769
Landrus, or Bear run, mine, Bloss coal bed at, ...2273
Lansdale shales described, New Red, ...2609
 "   " fossils, New Red. 2610
 "   " soil, New Red, ...2626
 "   " trap, New Red. 2619,2623

Larimer mine, Pittsburgh coal at, ...2511,252J
Larmon opening, bed D at, ...2252
Lattimer colliery, Eastern Middle coal field, ...2031
Lauder basin, Kelley seam in, Broad Top coal field, ...2173
Laughlin and Jones well, thickness of Venango group in, ...1543
Laurel Hill, Mauch Chunk No. XI rocks along, ...1839
" " anticlinal axis, ...1413,1697,1842,1861,1889,2219,2221,2232,2318
" " gap, 1573; quarry, 580; Slate Company's quarry, No. III
      slate, ... 609
" " gap, Pocono rocks in, ...1709
" " mine, Pittsburgh coal bed at, ...2541,2543
" ridge, ...1844,1845,1852,1892
Laurentian country, trap dykes in, Lake Superior to Labrador, ... 97
" mountains of Canada, ...61,82
Lawrence colliery, Western Middle coal field, ...2003
" county, No. XII in, ...1909
" " No. XIII in, ...2391
" " No. XIV in, ...2439
Law's, T. R. mine, Washington coal bed at, ...2576
Layton Station wells, Venango oil group in, ...1543
Lead in the New Red of Bucks and Montgomery counties, ...2631
" ore veins in formation No. II, Chapter XXXVII, 436; in No. IV,
    678; Selinsgrove, ... 962
Leathercracker limonites, 253; anticlinal and cove ores, ... 409
Lebanon, 270, 273, 288, 309, 312; level in, 274; level of divide near, ... 272
" county, Cornwall, 915; Cornwall magnetic iron ore mines, 1387
" " No. VI in, Chapter LXVII, ... 936
Ledger vein colliery, ...2100
Leechburg (Apollo) synclinal, Freeport group in, Jefferson co., ..2295,2300
" synclinal, Pittsburgh coal in, Allegheny county, ...2525
Lee colliery, Northern coal field, ...2003,2006
Leggett's gap, green beds of No. XI in, ...1827
Lehigh Coal and Navigation Company, coal lands in Panther creek
    basin, ...2086
" region, area of coal in, ...2151
" Valley Coal Company, York farm colliery, Southern coal field,2107
" " " " collieries, Northern coal field, ...1995,1997
" and Wilkes-Barre Coal Company's collieries, Northern coal
    field, ...1995,1997
" county, geology of, 552; Great valley in, 341; maps, 234, 275;
    range of mountains in, 710; Saucon zinc mines, 436,
    439; slates, 563, 564; slate quarries, 574, 588; South
    mountains of, ... 61
" region, 299, 298; survey, 345; No. II in the, ... 301

INDEX FINAL SUMMARY REPORT. 53

Lehigh river, hydraulic cement quarries, Chapter XXIX, 337; level of the bed of, at Allentown, 434; No. VII in the Delaware and, Chapter LXXVI, 1045; No. VIIIb, Marcellus at the, 1211; No. IX Catskill on the, 1594; section of No. III, slates, 575, 1590; section down the, from Mt. Pisgah to Lehighton, 1595; sections between the Delaware and, 1598; White's section on the, Nos. IX and X, ...1596,1598
" slate belt, 556; roofing slate region. 139
" Water gap, 560, 576, 586, 587, 631, 675, 710, 726, 729, 734, 735, 756, 864, 890, 909, 917, 920, 922, 925, 929, 931, 935, 936, 937, 1037, 1043, 1045, 1047, 1063, 1073, 1081, 1084, 1093, 1175, 1176, 1178, 1180, 1184, 1182, 1211; (measurements of No. IV), 641, 674; (measurements of No. V), 731; No. VI at the, 921; paint mines and cement quarries of the, ...1067
Lehighton, 1327, 1328, 1594; anticlinal, 1569; section from Mt. Pisgah to, ...1595
Lehman, Mr. A. E. topographical map of South mountains............ 145
Leiper bank, Pittsburgh coal bed at, ...2539
" granite quarries, ... 97
Leisenring Nos. 1 and 2 mines, Pittsburgh bed at. 2492
Leith mine, Pittsburgh bed section at, ...2495
Lelar coal bed, Southern coal field, ...2084
Lemon opening, Pittsburgh bed at, ...2488
Lemont furnace opening, Pittsburgh bed at. 2494
LeRoy (Gulf Brook) section, A. T. Lilley, ...1448,1453,1456,1457,1461,1462,1466
" quarries, Onondaga limestone in. 1145; quarry section. 1159
Lesley J. P. report to Lyon, Shorb & Co., limonite ores................ 378
Lesquereux, Leo palaeozoic fossils in the New Red. 2613
Levan's drift, Southern coal field, ...2088
Lewis bank, bed D at, ...2342
" coal bed. "Peach mountain bed." Southern coal field, ...2104,2115
" colliery, Southern coal field. 2114,2115
Lewis' quarry, Hamilton fossils at. 1283, 1295; Venango 3d S. S. and Panama conglomerate, ...1517,1521
Lewisburg, exposures at, Bloomsburg red shale of No. V. 872
" iron furnace quarries. limestone of No. VI. 954
Lewistown, No. V at. 791
" section, 781, 1027, 1093, 1225; Logan gap section. 885
" synclinal. Marcellus ore of Huntingdon county. 1229
Lickel's ore bank, Upper Hamilton fossil ore. 1268
Ligonier basin. 1889,2318,2362.2450,2473
" " No. XIII in, ...2317
" valley. 1699,1701,1847,1852,1891
" " Indiana county. No. XIV in. 2424
" " Westmoreland county. No. XV in. 2470

Lilley, Mr. A. T. 1438, 1456; section at LeRoy (Gulf Brook),..........1461,1466
Lima and Findlay, Ohio, oil and gas belts, ... 493
Lime Ridge quarries, bastard beds of Bossardville limestone, ...948,1025
" and its use upon land, ... 976
Limestone beds in Northern coal field, 2020; coves in slates of Great
  valley, 283; No. III, Washington county group, No. XVI,
  2574; No. V, Washington county group, No. XVI, 2572;
  quarries, Great valley, east of Susquehanna, Chapter
  XXVI, 309; quarries, Great valley, west of Susquehanna,
  Chapter XXVII. 324; Ridge, No. VI quarries in, 955; Ridge,
  section of No. VI, 944; white in No. II, Chapter XXXIX,... 467
Limonites of No. II, Centre county, Chapter XXXII, 372; in Primal
  upper slate, Chapter XX, 205; precipitated from pyrites,
  435; precipitated in caves, ... 433
Lincoln colliery, Southern coal field, ...2127,2128,2129
Linton's hill, 1560; sections near, ...1558
Lippincott mine, Pittsburgh coal at, ...2486
Lisbon basin, No. XIII in, ...2300,2306
" (Irwin) gas coal basin, No. XV in, ...2434,2469,2501,2515,2565
Listie mine, bed C' section at, ...2252
Little Alps mine, Pittsburgh coal at, 2521; Aughwick valley, 1222;
  Black creek coal basin, 2030; Bushkill creek, 1249, 1252; falls.
  1317; Clinton coal bed, Southern coal field,.2114; coal bed, No.
  6 or Lykens Valley bed, Southern coal field, 2135; Diamond
  coal bed, Southern coal field, 2113, 2123; Diamond coal bed,
  Western Middle coal field, 2062, 2071; Fishing creek, 739, 949.
  1071, 1074, 1133, 1261, 1265, 1564; gap, 370; section, 1266; Germany,
  979, 1213, 1217, 1271, 1275; fault, 979; Horseshoe dyke, trap in No.
  II, 458, 459, 460,463; Illinois valley, No. IV in, 682; section of Clinton rocks in, 757, 758; Juniata river section, No. II in, 406; Orchard coal bed in Southern coal field, 2102, 2113, 2123; Orchard
  coal bed in Western Middle coal field, 2058, 2062, 2066, 2071; Pine
  creek basin, Lycoming county, No. XII and No. XIII in, 1865.
  2185; Pine creek coal basin, description of, 2188, 2189; Pittsburgh mine, Pittsburgh coal bed at. 2519; Tracy coal bed in
  Southern coal field, 2103, 2114, 2124; Tracy coal bed in Western
  Middle coal field, ...2058,2062,2066
Llewellyn-Tremont division, ...2117
Lloydsville mine, bed B in, ...2231
Lock (Canoe) mountain, 404, 408, 409; anticlinal, ... 414
Lockhart mine, Pittsburgh coal in, ...2550,2551
Lock Haven, gap at, No. IV at, ... 667
  "   "   long section, Rep. G4, ... 489
  "   "   section, Mill Hall part of, ... 863
  "   "   -Williamsport valley, Oriskany and Caudagalli at, ...1143
  " Ridge, southern range of limonite ore, ... 232

Locust Gap colliery, Western Middle coal field, ...2060,2061
" mountain, 2049; drifts, 2087; gap, ...2097,2098,2102
" run colliery, Western Middle coal field, ...2059
" Spring colliery, Western Middle coal field, ...2060,2061
Logan colliery, Western Middle coal field, ...2065
Logan, Sir William, Huronian system on Lake Huron, 152, 153; magnetic ore mine, York county, 258; gap, No. III in, 551, 560; No. IV in, 421, 631, 645, 651; gap, No. V in, 726, 727, 787, 791; limestone and Salina variegated shales, ... 805
" section of Clinton No. V rocks, 793, 797; of Salina and Oneida rocks, 787; of Bloomsburg red shales, 790; Salina upper shale, 791; of fossil ore beds of No. V, 728; of ore sandstone of No. V, 803; thickness between cre S. S. and Oriskany S. S., 804; upper Olive shales No. V, 807; middle Olive shales, 814; iron sandstone group of No. V, ... 815
Loganville, York county, pryllite beds, 134; dyke, ... 454
" iron ore deposits near, ...136,212,221,222
Lomison, Dr. pit, Pittsburgh coal at, ... 2485
Long, C. C. bank, Pittsburgh coal at, ...2485
Long Hollow, 1096; cross section, ...1097
" Narrows, description, 729, 783; section of No. V in, ... 785
Longwood Coal Company's mine, section and Alton coal at, ...1880
Loop mountain, contact of No. II and No. III, ... 363
" " iron ore pits in No. V, ... 854
" run country bank drift, description of, Tipton run, ...1687,1689
" " drift, Ashburner's section of, coals of No. X, ...1683
Loose's quarries, No. II limestone in, ... 314
Lorberry gap, No. XII and Lykens Valley beds, ...2140,2141
Lost Creek Valley, limestone beds at, 806; ridge, ...807,814,815,817
Love opening, Pittsburgh coal at, ...2488
Lovedale mine, Pittsburgh coal at, ...2534
Lovetown and Warrior Mark iron ore range, Centre county limonites, 371
Low Brothers' quarry, fossil beds at, Bossardville limestone, ... 951
" quarry, 939; section of No. VI, analyses, ... 941
Lower Harmony quarries, Warren county, N. J., marble beds at, ... 109
" Helderberg No. VI in Mohawk valley, 904; Northern and Eastern outcrops in New York, 902 to 913; in Eastern Pennsylvania, Chapter LXVI, 917; in Carbon, Schuylkill, Lebanon and Dauphin counties, Chapter LXVII, 936; on North branch of Susquehanna river, Chapter LXVIII, 940; at Selinsgrove, on Susquehanna, Chapter, LXIX, 957 in Perry county, Lower Juniata, Chapter LXX, 967; sub-divisions in Perry county, 970; in

Union, Snyder, Mifflin, Huntingdon, Bedford and Fulton counties, Chapter LXXI 9S3; in Blair, Centre, Clinton and Lycoming counties along the Bald Eagle valley, Chapter LXXII, 999; lead and zinc ores, 962; limestone boulder slabs, 977; folded structure, 978; quarries in Perry county, 980; quarries in Columbia county, 949; in Montour and Northumberland counties, 953; bone cave in, 916; Gypsum beds in New York, 913; fishes, 775; fossils, Chapter LXXIII, 1005; sponge coral reefs in, Chapter LXXIV, ...1020
Lower Juniata, No. VI in Perry county on the, Chapter LXX,........ 967
" " No. VII on the, Chapter LXXVIII, ...1084
" " river, 753, 755, 1328; region, ... 893
" Mt. Bethel township, Northampton county slate quarries of No. III,, ... 592
" Productive coal measures, unique features of in Cambria co., 2220
Lowry and Eschelberger's quarry, oxide of manganese at, ...1801
" opening, bed E in, ...2325
Loyalhanna Coal Company's shaft, Pittsburgh coal at, ...2483
" river at Ursina gap, Catskill rocks, ...1573
Loyalsock and Mehoopany coal field, description of. 2007
Lucy Furnace Company quarry, No. VI limestone in Mifflin county,... 989
Ludlow oil well, thickness of Upper Chemung, ...1482
" (English) fish beds, ...770-771
Ludwig's quarry, No. II limestone in Berks county, ... 312
Luke Fidler colliery, Western Middle coal field, Lykens Valley and Holmes beds at, ...2069,2071
Lusk's bank, Mercer lower coal at, ...1901
Luther's well, gas vein in No. VIIIf, Erie county, ...1354
Luzerne county, Coxton section of Catskill rocks in, ...1607
Lycoming county, thickness of No. IV in gaps of. 631, 649, 667; No. VI in, Chapter LXXII, 999 and 1004; No. VII in, 1123; fish beds in, 1466; No. IX Catskill in, 1621; sections of No. XI in, 1833 and 1835; No. XII in. 1864 and 1865; Siliceous limestone in, 1801; No. XIII in. 2185
Lykens (mine No. 2) mine, 407; pits, limonite ores of No. II, ... 407
" Valley coal beds in the Southern field, 2080, 2083, 2087, 2095, 2108, 2119, 2127, 2135, 2142; beds in the Western Middle field, 2051, 2055, 2059, 2064, 2068; specific gravity of, ...:......1929

Lykens Valley No. 1 bed, Lincoln colliery, Southern coal field, 2129, 2136; No. 1½ bed, Lincoln colliery, Southern coal field, 2129; No. 2 bed, Southern coal field, 2128, 2136; No. 3 bed, Southern coal field, 2128, 2136; No. 4 bed, Southern coal field, 2128, 2136, 2142; No. 5 bed, Southern coal field, 2119, 2128, 2136, 2142; No. 6 bed, Southern coal field, 2127, 2135, 2142; tunnel, Mammoth bed in Southern coal field, ... 2137
Lyle mine, Mercer upper coal at the, ... 1903
Lyman, B. S. 1220, 2078; survey, 48; sections, ... 179,324
Lyman camp mine, Alton lower coal at, ... 1877
Lynch opening, Pittsburgh coal at, ... 2520
Lynn township, Lehigh county, quarries, roofing slates of No. III... 609
Lynnport quarry, No. III slates in, ... 580,609
Lyon, W. M. mine, Pittsburgh coal at, ... 2521
Lytles pit, Pittsburgh coal at, ... 2529

M.

Madisonburg gap cross section, No. II and No. III rocks in, ... 369,370
Maffet colliery, Northern coal field, ... 1989
Magnetic limonites, either Primal, Gneiss or Trias, Chapter XXI... 256
Mahanoy coal basin, general description of, in Western Middle coal field, ... 2052
" ridge basin, fossils in Marcellus formation, ... 1215
Mahoning sandstone, ... 1713,2156,2169,2175,2220,2287,2402,2408
" " section of, first basin, Cambria county, ... 2413
Maltby colliery, Northern coal field, ... 1990,1992
Mammoth coal bed, 1737; in the Broad mountain basins, 2079; Eastern Middle field, 2030, 2031, 2033, 2035, 2036, 2037, 2040, 2043, 2046, 2047; Heckscherville basin, 2084; Panther creek basin, 2089; Southern field, 2098, 2110, 2121, 2130, 2137; Western Middle field, 2056,2061,2065,2070
" Nos. 1 and 2 mines, Pittsburgh coal at, ... 2488
Manayunk Mica Schist gneiss group, ... 122,123,125
Mannigton well section, mountain limestone and No. X rocks, ... 1715
Manor Gas Coal Company's opening, Pittsburgh coal at, ... 2510,2511
Mansfield Coal and Coke Company's mine, Pittsburgh bed at 2542
" mine, Pittsburgh coal at, ... 2541
Manville colliery, Northern coal field, ... 1970
Maple Glen mine, Pittsburgh coal bed at, ... 2558
Maple Grove Coal Company, Maple Grove coal at, in Mercer county, 1903
Mapleton glass sand quarries, ... 1763
" section, ... 1278,1313
Mapletown bank, Sewickley coal bed at, ... 2562
Marble Hall quarries, white limestones and marbles of No. II, ... 471

Marble mountain mine, red hematite in talcose formation, N. J., ... 116
" white in No. II, Chapter XXXIX, 467; black in No. II, Chapter XL, ... 479
Marcellus No. VIIIb Chapters LXXXIV to LXXXVI, 1194; in Pennsylvania, Chapter LXXXV, 1203; in Perry county, 1211; at the Lehigh, 1211; in Middle and Eastern New York, 1200 and 1201; false coal beds, 1217; fossils in Western New York, 1198; fossils in Middle New York, 1200; iron ore, Chapter LXXXVI, 1220; in Mifflin county, 1223; in Huntingdon county, 1228; at Orbisonia, 1231; in Juniata county, 1221; in Bald Eagle Valley, 1234
Marcy coal bed, in Scranton division of Northern coal field, ...1980
Margie Franklin colliery, Western Middle coal field, ...2069
Marion basin, Indiana county, general description of, ...2336
Marshall mine, Pittsburgh coal bed in, ...2528
Marshburg bed, sections of in Cameron county, ...1881
" lower coal in McKean county, ...1737,1840
" upper coal, in McKean county, general description of, ...1877
Marshfield colliery, Southern coal field, ...2125
Marshwood rock slope, Northern coal field, ...1961
Martin's mine, bed D at, ...2246
" pit, Pittsburgh coal bed in, ...2558
Martinsburg limestone, description of, in Washington county, ...2579
Marvin creek limestone, in McKean county, ...1731
Maryland State line, No. VI on the, ... 987
Masillon sandstone of Ohio, ...1905
Mason's pit, Pittsburgh coal bed in, ...2529
Masontown coal, ...2408
Massachusetts New Red, description of, ...2600,2611,2614,2615
Mather's farm, quarry on, limestone No. II, ... 470
Mathew opening, Freeport lower coal D, Jefferson county, ...2301
Matilda furnace section, Danville ore beds in, ... 812
Mauch Chunk formation in Clearfield county, 1840; gap, Pocono S. S. in, 1723; lower shales, on Conemaugh and Loyalhanna creeks, 1846; mountain, exposures of Nos. VIII, IX and X, 1504; outcrop geography of, 1809; red shale, 1681, 1719, 1751, 1783, 1785, 1789, 1791, 1803, 1850, 1858, 1866, 2166, 2220, 2292; red shale around Northern coal field, 1823; red shale, general description of, 1815; red shale, No. XI, on Salt Lick creek, Cambria county, 2237; red shale No. XI, Chapter CXV, 1805; red shale, Westmoreland and Fayette counties, 1843; rocks No. XI, Somerset county, 1842; section, Blair county, 1841; series, in Northwestern Pennsylvania, 1835; in Broad Top region, 1835; and Pocono rocks, absence of conglomerate beds in, ...1647

Mauser's quarry, section of No. VI, ...939,941
Maxwell colliery, Northern coal field, ...1992
McCartney's land, siliceous limestone on, ...1843
McCauley basin, Eastern Middle coal field, ...2034,2037
McClary's opening, Waynesburg coal bed in, ...2561
" quarry, Connoquenessing S. S., Mercer county, ...1903
McClelland mine, bed E in, ...2346
McClintock mine, bed C' in, ...2262
McClure, W. pit, Waynesburg coal bed in, ...2563
McConnellsburg cove, limestone and ores, Chapter XXXV, 423; mountains of No. IV in, ...683
McCormick mine, Pittsburgh coal bed in, ...2544
" quarry, limestone of No. 11 in, ...322,325
McCrea furnace, lower productive coal series, Armstrong county, ...1890
McCreary mine, analysis of Pittsburgh coal from, ...2487
McCree's opening, Washington coal bed in, ...2575
McDonald's slope, Southern coal field, ...2122
McGary E. opening, bed D in, ...2301
McIntyre basin, No. XII in, ...1871
" coal bed section, ...2186
" mine, bed D in, 2186; bed E in, ...2345
" opening, Barnet coal bed in, ...2172
" -Ralston basin, ...2185
McKean county, No. XII series in, groups XII and XIII, 1866, 1873; Bradford oil field in, 1484; Chemung in, 1481; map, 1492; No. XII in, 1858; No. XIII in, 2277; oil sands, 1486, 1489; Wilcox gas wells sections, ...1485
McKee mine, bed D in, ...2302
McKee's gap, Danville ore bed, ...854,856
McKee mine, Pittsburgh coal bed in, ...2547
McLean opening, bed B in, ...2335
McMaster bank, bed D in, ...2323
McMillan and Hopkins mines, Pittsburgh coal bed in, ...2518
McMillan's opening, bed E in, ...2397
McVeytown, section of Oriskany S. S. at, ...1093 to 1096
Meade run coal basin, No. XIII in Elk county, ...2287
Meadow Brook colliery, Northern coal field, ...1969
Meadville group of Prof. White, Pocono S. S. No. X, ...1757
" limestone, ...1731
" lower limestone, ...1763
" lower shales, ...1761
" quarries, ...1900
" upper limestone, ...1759
Mechanical deposits of No. II, Chapter XLIII, ...497
Medina model of upper surface, 703; fossils, Chapter LIV, ...714

Mehoopany coal basins, Wyoming county, ...2013
" No. 1 well record, ...1582
Mercer's coal bank, Mercer coal bed in, ...1904
Mercer coal bed 1863; coal group, 1860, 1869, 1887; county, No. XII in, 1902; county, No. XIII in, 2387; group, 1897; Iron and Coal Company, Quakertown coal bed, 1906; limestones, 1914; lower coal, 1904, 1911; lower iron ore, 1903; lower limestone, 1904; lower shales, 1905; shales, 1903; upper coal, 1910; upper limestone and coal, ...1902
Merchant mine, Pittsburgh coal bed in, ...2521
Merriam colliery, Western Middle coal field, ...2060,2061
Mesozoic age and sub-divisions, ...48
Meyers mine, bed C in, 2250; Waynesburg coal bed in,...............2556
Mid Cove dyke, trap dykes of No. II, West Duncannon dyke, ...458,460
Middle Anthracite basin, ...706
"   creek colliery, Southern coal field, ...2118,2120,2121,2122,2123
Midvalley No. 1 colliery, Western Middle coal field, ...2065,2066
"   No. 2 colliery, Western Middle coal field, ...2065
Midway mine, Pittsburgh coal bed in, ...2541,2544
Mifflin county Bloomsburg red shale in, 790; Logan section, Clinton rocks, 793 to 797; Logan gap section, 885; measurement of No. IV, 631, 645; No. VI in, Chapter LXXI, 983, 987 to 989; No. VII in, 1093, 1097; No. VIIIc in, ...1276
Milesburg gap, Clinton olive shales in, ...861
Milesville mine, Pittsburgh coal bed in, ...2538
Milford colliery, Southern coal field, ...2103,2104
"   N. J. fossils in the New Red, ...2609
"   tunnel, diamond bed? in, Southern coal field, ...2103
Milton anticlinal, ...955,1261,1329
Mill creek gap in Mine Hill, "Scoot Steel Coal," Southern coal field,...2083
"   "   -Pine creek basin, description of, in Potter county, ...2276
"   Hall gap, No. IV in, 647 and 667; No. V in, ...862
"   Hollow colliery, Northern coal field, ...1988,1990,1992
Miller bank, Brush creek coal bed in, ...2440
"   J. M. mine, Waynesburg coal bed in, ...2557
"   mine, Pittsburgh coal bed in, ...2463
"   bank, Pittsburgh coal bed in, ...2503
"   "   bed D section at, ...2217
Millertown, section of Clinton rocks at, ...757
Milligan's cove ore and limestone, Chapter XXV, 420; measurements of No. IV in Will's mountain gap, ...649
Mills coal bed, Orchard Slope No. 4, or "H bed," Northern coal field, 2006
Millwood shaft, rock fault in, Westmoreland county, ...2483
Mine Hill colliery, Southern coal field, ...2085
"   "   or Heckscherville valley basin, ...2081
Mingot mine, Pittsburgh coal bed in, ...2552

INDEX FINAL SUMMARY REPORT. 61

Mining methods in the anthracite region, ...1932
Missouri (limestone caverns, etc.), ...438
Mitchell mine, Bloss coal bed in, ...2274
Metz's quarries, No. II limestone, Dauphin county, ...322
Monaghan's pit, Pittsburgh coal bed in, ...2524
Monastery lands, Alton coal beds on, Cameron county, ...1883
"   " coal openings, Dagus coal bed in, ...2284
" mine, Pittsburgh coal bed in, ...2485
Monitor colliery, Southern coal field, ...2060,2112
Monongahela Navigation Company, table of shipments of coal, ...2526
Monroe colliery, Western Middle coal field, ...2065
" county, Catskill rocks No. IX in, 1585; limestone quarries, Bossardville No. VI limestone, 929; Delaware flagstone beds in, 1591; Catskill outliers, ...1574,1575
Montebello gap section, Perry county, Hamilton rocks, ...1271
Montgomery county (limestone basin), marble, 468; limestone valley, primal slates and sandstones in, 183; (marble quarries), 471; apatite crystals in Prince's quarry, 114; (White limestones and marbles of No. II in Chester, York and Centre counties) Chapter XXXIX, ...467
" pit, Pittsburgh coal bed in, ...2558
Montmorency basin, No. XIII coal measures in and sections, ...2290
Montour anticlinal, exposures of No. VI upon, 937, 1071; No. XI, 1173; mine, Pittsburgh coal bed in, 2544; region, quarries of No. VI, 890; ridge, No. V in, 737; ridge (No. VII around), Chapter LXXVII, 1071, 1073; ridge, No. VIIIc, Hamilton, north of, 1264; ridge, structure, ...941 to 948
Montrose group, No. IX Catskill rocks, fossils in, ...1608
Montville, limestone type of rocks, New Jersey Archaean, ...72
Moose Head, section of Mauch Chunk No. XI shale at, ...1819
Moosic mountain, Griswold conglomerate in, ...1647,1649
Morea colliery, Southern coal field, ...2079
Morgan mine, bed B in, ...2263
Morgantown sandstone of No. XIV described, 2407; over Murraysville axis, ...2532
Morrell mine, Pittsburgh coal bed section at, ...2492
Morris Coal Company's shaft, Sharon plant shales at, ...1906
" pit, bed D in, ...2301
" ridge, colliery, Western Middle coal field, ...2065
" run mines, Seymour and Bloss coal bed at, ...2271
Morrison's Cove and Canoe Valley limestone and ore, Chapter XXXIV, ...401 to 418
Moshannon bed D, in Clearfield county, described, ...2210
" mine, bed D in, ...2236
Mosier's drift, Southern coal field, ...2083

Mosquito valley, No. II limestone exposed in, ... 365
Mount Eagle road, coal beds along, ...2143
" Hope, type of New Jersey archaean described. 71,72
" Pleasant conglomerate, description of, ...1647
" Union section, Nos. V, VI and VII, ... 799
Mountain Creek, valley and limonite banks, ... 241
" " valley, quarries in, in No. II, ... 324
" limestone, Chapter CXIV, ...1633,1793,1846,1847
" " (No. X-XI) Chapter CXIV, ...1789
" " section of in Ligonier valley, ...1848
" " in Lycoming county, ...1801
" " on Trough creek, ...1799
Mountains of Pennsylvania, names of, 681; parallelism, 686; convergence, 688; spurs, 689; crests, single and double, 695; difference in height, 696; Keel mountains of, 697; anticlinal vaults restored. 699
Moyer's quarry, limestone of No. II, Dauphin county, ... 319
Mt. Alto quarry, limestone of No. II, Franklin county, 326; Carmel colliery, Western Middle coal field, 2065, 2066; Equity basin, Broad Top coal field, 2172; Holly cross section, thickness of felspathic felsite series, 147; Holly range, South mountains, 143, 146, 150, 167; Hope mine, Clarion coal A' in, 2280; Hope slope, Southern coal field, 2110; Jessup colliery, Northern coal field, 1963; Morris coal shaft, Sharon coal bed in, 1907; Morris well, section in, at Pittsburgh, 1715; Pisgah, section from, over Mauch Chunk down the Lehigh, etc., 1595; Pisgah, Winslow's section of, 1813, 1875; Pleasant colliery, Northern coal field, 1970, 1971; Eastern Middle coal field, 2039; Pleasant section, No. VIII rocks, 1266; Savage coal bed, 1864, 1991; Savage fire clay analysis, 2244; bed, 2243; Savage, group in the Broad Top basin, 1863; Union (measurements of No. V), 803, 804; Union, section, No. VI rocks, 1027; Union section, composition of the, 803; Vernon colliery, Northern coal field, ...1961
Mulbaugh Hill, Archaean rocks, 70; Potsdam S. S. in, 181; described...291
Mull mine, bed E at, ...2261
Muncy Hill range, No. VIII rocks in, ... 875
" section of No. V, ... 866
Murphy's opening, Pittsburgh coal bed at, ...2484
Mutual, Nos. 1, 2 and 3 mines, Pittsburgh coal bed at, ...2488

N.

Names of the formations, Chapter VI, ... 39
Nanticoke collieries, Northern coal field, ...2003,2004,2005,2006,2007
" gap, section at, Nos. X, XI and XII, ...1827
" Hill, bore hole No. 2, in Northern coal field, ...2002

Nanticoke Shickshinny division of Northern coal field, ...1999
Natalie colliery, Western Middle coal field, ...2065
National Coal Company's mine, lower division of Pittsburgh bed at, 2541
" colliery, Northern coal field, ...1969
" quarry, roofing slate of No. III, ...691
Nearpass' quarry, Stormville limestone of New Jersey, 924; No. VI
section at, ...919
Negro mountain, ...1697,1870,2244
"   "   anticlinal, ...1842,2241,2246,2248,2252,2461
Neil's colliery, Southern coal field, ...2113
Neilson colliery, Western Middle coal field, ...2071
Neleigh's cut, section of Mauch Chunk red shale in, ...1817
Nellie mine, Pittsburgh coal bed in, ...2491
Nescopec anticlinal of Luzerne county, ...706
" gap, section of VIII to IX (Catawissa), ...1604
Neshannock Coal Company's shafts, Sharon coal bed in, ...1908
Nesquehoning or Broad mountain, described, No. X, 1653; No. XI,..1813
" railroad tunnel, "A" coal in, Southern coal field, B.
& C., ...2058
Neville shaft, Southern coal field, ...2084
New Bangor quarries, roofing slates of No. III, ...534
Newark and Belleville, N. J., fossils of New Red at, ...2609,2613
" system, name, described, ...2589
New Boston colliery, Southern coal field, 2079; Boston-Gordon coal
basin, Southern coal field, 2079; Britain coal in the New Red,
2629; Catsburg mine, Pittsburgh coal bed in, 2552; coal bed,
Northern coal field, 1998; Coal Bluff mine, Pittsburgh coal bed
in, ...2552
Newcomer opening, Pittsburgh coal bed in, ...2490
New county coal bed, Northern coal field, 1970; Eagle mine, Pittsburgh coal bed in, 2552; England (glacial covering), 51; Galena
lead mine, Doan's mine in New Red, 2631; Hampshire, geological section across, 58; Hampshire slates, 585; Hope trap, ...2626
New Jersey, Archaean types in, 71; azoic areas, 162; Big mine in Sussex county, 442; (flagstone quarries), 569; Franklin zinc
mine, 436; geology of, 676; geological map and geology
of, 109; Highland mine, 115; (limestone) analyses, 479;
(limestone) valleys of Northern, surveys of, 474; (limonite mines), 252; magnetic iron ore mines of Northern,
447; marble beds, 109; (Medina-Oneida mountain range),
709; New Red, 2600, 2610, 2611, 2615, 2616, 2621, 2623; No. IV
in, 676; Pequest belt, 116; (slate belts), 583; (slates of the
Great valley, in New York and Eastern Pennsylvania),
562; slate quarries in the Great valley of, 574; (white
marbles), 481; (zinc), 438; and Pennsylvania Archaean
Highland belt of, ...63

Newkirk colliery, Southern coal field, ...2096,2097,2098,2101,2102
New Lincoln colliery, Southern coal field, ...2127,2128,2129
" Red, name discussed, ...**2589**
"   "   palaeontological age of, ...**2615**
Newville Slate Company's quarry, roofing slate of III, ... 601
New York, Adirondack mountains, in Northern, 671; black marble at Glen Falls, Northern, 482, 483; black marble in No. VI, 484; Chemung in western, 1374; formations, names of, 42; geology of, 24; geology of southern, 40; mountain range, (Medina-Oneida), 709; New Red in, 2615; (nomenclature), 42, 47, 301, 299; (nomenclature and divisions of the Hamilton sandstone and shale), 1236, 1237; No. IV in, 677; State, No. VIIIc Hamilton in, 1237 to 1246; No. VIIId Tully limestone in, Chapter XC, 1307; Panama conglomerate of, 1519; Prosser's general section of Western Middle, 649; well records, Prosser and Ashburner, 1191; slate belt in Washington county, Rensselaer county and Columbia county, 583; slates of the Great valley in, New Jersey and Eastern Pennsylvania, 562; and Pennsylvania quarry, No. III roofing slate, ... 601
Niagara No. V, Chapters LV to LXIII, 721; in Pennsylvania, evidence of, 883 to 889; not found on Juniata, ...746,760
Nippenose valley, Trenton shells in, 523; slate rings of, ... 535
Nittany valley limestone cross sections, Chapter XXXI, ...365,535
"   "   ore mines, ... 378
"   "   ore, Chapter XXXIII, ... 387
Nockamixon township copper, New Red formation, ...2632
Noel's bank, Pittsburgh coal in, ...2485
Nolo axis, Freeport coal E over, Indiana county, ...2320
Nolte mine, Pittsburgh coal in, ...2544
No. VI in Eastern Pennsylvania, Chapter LXVI, ... 917
Nomenclature, Chapter VI, ... 39
Norristown shales described, ...2612
"   "   fossils, ...2612
"   "   palaeontological age, ...2614
"   "   soil, ...2626
"   "   trap, ...2619,2623
North, or Allegheny mountain, ...1571
Northampton county, belt of quarries across, roofing slates No. III, 578
"   "   upper limit of No. III slate in, ... 710
"   serpentine, 105; serpentine beds north of Easton,... 107
"   "   slate quarries, ... **574**
"   quarry, No. III slate, ... 598
"   and Lehigh counties, roofing slate belt, ...: 141
"   "   "   "   slate quarries, Chapter XLIX, ... 583

North Bangor quarry, roofing slate No. III, 584; Branch Susquehanna river, No. V on the, Chapter LVI, 737; No. VI on the, Chapter LXVIII, 939; No. VIIIa on the, 1180; No. IX Catskill on the, 1604; Brookside colliery, Southern coal field, 2129; Carolina New Red, 2611, 2613, 2615, 2629; Valley Hill, quartzite or sandstone, 79; azoic hydromica slate in, 133; rock or Hellam quartzite, 166; upper primal slates in, 168; Potsdam No. I in, 173, 175, 176, 177, 178; vein drift, Southern coal field, 2142; Whitehall township quarries, Lehigh county, roofing slate No. III, ... 608
Northdale tunnel, Southern coal field, ...2097,2099
Northern anthracite basin, 706; Central Railroad, section along, Hamilton rocks, 1255; coal field, contents of, 2147; coal field, its size and location, 1946; coal field, thickness of coal measures in, 1951; field, Pottsville conglomerate in, 1854; field, structure of, ...1947
North Franklin colliery, Western Middle coal field, 2067; Franklin No. 1 colliery, Western Middle coal field, 2069; Garden, Nevin's limestone quarry in, apatite, 114; mountain, mineral worthlessness of the, along the Great Valley, and of all the other mountains of IV in Middle Pennsylvania, Chapter LIII, 712; Peach Bottom Company's quarry, quartz between slate laminae at, 568; Point basin, Kelly coal in, Broad Top coal field, 2173; Savage Fire Brick Company's mine section, 2243; Stars mines, upper Freeport coal E in, ...2348
Northrop mine, bed B in, Barclay basin, ...2268
Northumberland county, No. V in Union, Snyder and, counties, ... 870
" " No. VI quarries in, ... 353
" " No. VIIIa in, ...1178
" synclinal, Bossardville No. VI limestone in, ... 985
Northwest colliery, Northern coal field, ...1955,1957
Northwestern Pennsylvania, No. XI in, ...1833
Norwich basin, Alton lower coal bed in, ...1877
" or Potato creek basin, described, 1875; anticlinal, ...1876
Nottingham colliery, Northern coal field, ...1983
" mines, Pittsburgh coal in, ...2551
Novelty slope, Southern coal field, ...2115
Nye's drift, Southern coal field, ...2132

O.

Oak Hill Coal Company's colliery, Mercer lower coal in, 1905; colliery, Southern coal field, 2105, 2111; Holmes bed in, 2112; Orchard bed, 2113; colliery, Northern coal field, 1980; collieries, 2528; mines Nos. 3 and 4, Pittsburgh coal at, 2528, 2530, 2531; Ridge colliery, bed D at, Cambria county, Hastings district, 2241; Ridge mine, Pittsburgh coal at, ...2541

66 GEOLOGICAL SURVEY OF PENNSYLVANIA.

Oakdale colliery, Primrose bed at, Southern coal field, ...2085
Oakgrove synclinal, red specular iron ore in, Perry county, ...1603
Oakland Coal Company mines, Sharon coal bed at, ...1906,1908
Oakwood colliery, Northern coal field, ...1987
"   " tunnel, Kelly coal in, 2176; section and Barnet bed, ...2178
"   " Nos. 2, 3 and 4, ...2535
Ocean mine, Freeport sandstone at, ...2176
O'Connor's, J. mine, Pittsburgh coal in, described, ...2576
Offset mountain, conformity of No. IV on No. III in, ... 710
Ohio (fossil fish in, carboniferous limestone), ... 773
" geology of, 24; geology of eastern, ... 40
" Mining Company's colliery, lower Freeport coal D at, ...2305
" oil and gas districts, Trenton No. II limestone, ... 495
" river salt region, blow wells in, ...1486
Ohlen wells, Nos. 1 to 4, Chap. XCIX, Nos. VIII and IX in, ...1471 to 79
Oil creek, oil sands on, described, 1483; creek lake bore hole record,
1510; creek lake group (White), description of, 1767; Creek Lake
Lumber and Mining Company, wells Nos. 1 and 2 record, 1507;
gas in No. II, Chapter XLII, 492; gas of Portage, 1352; sands of
Chemung in McKean county, Chapter C, 1481; sands of Warren
county, Chapter CI, 1489; sands of Bradford and Elk, ...1484,1485
Old Bennett coal bed, Old Orchard bed in Northern coal field, ...1994
" Forge colliery, Northern coal field, ...1997
" Seat bank geological range of pipe ore horizons at, ... 398
Older Mesozoic, name, ...2589,2590
Olean Conglomerate,1721,1733,1753,1781,1839,1861,1873,1892 to 1894,1900,1901,2277
" rock, description of, ...1874
" (sub) Shenango shales, Warren county, description of, ...1737
Olyphant colliery, Northern coal field, ...1965
" Nos. 1 and 2 coal bed, Northern coal field, ...1965,1973
Oneida colliery, Eastern Middle coal field, ...2045,2046
" falls, Pentamerus limestone at, 912; lower lime shales (Mar-
" cellus), ...1200
" fossils, Chapter LIV, ... 711
Oneonta sandstone, ...1365
Onondaga limestone, ...1143
O'Neill mine, Pittsburgh coal bed in, ...2534
Ontario anticlinal, No. XII outcrop shaped by, Northern coal field,...1975
" colliery, Northern coal field, ...1962,1978
Orangeville shales, description of, ...1765
Orbisonia (fossils found at and near), 1005, 1007, 1100, 1101; gap, thickness of No. II limestone at, 422; Marcellus ore at, 1231;
measured section through, 1231; measurements of No. V,
803, 804; mines Nos. 1, 2, 3, sections of No. V Ore S. S. at,
813; No. IV at, 653; range, condition of the Sand Vein ore
along, 829; section of Nos. VI and VII, 1027; Marcellus ore

beds in, 1229; section, composition of the, Nos. VII to No.
IV, ...799,801,802,803
Orchard coal bed, Northern coal field, ...1994,2006
" " " in Southern field, ...2102,2113,2123,2132,2137
" " " in Western Middle field, ...2057,2061,2066,2071
Ores of lead, zinc and copper, age, New Red formation, ...2632
Ore sandstone, a fixed horizon, 803; group, 808; in Huntingdon valley, 827
Original Petroleum Company below Watson's falls (well record), ...1539
Oriskany falls, Oriskany sandstone named from, ...1035
" No. VII, Chapters LXXV to LXXXI, 1034; in Union and Snyder counties, 1077; around Montour's Ridge, 1073; in the four counties, 1078; on Selinsgrove anticlinal, 1079; in iron ore in Perry county, 1090; in Juniata county, 1090; in Mifflin and southern Huntingdon counties, Chapter LXXIX, 1093; on upper Juniata, in Huntingdon, Bedford, Blair and Centre counties, Chapter LXXX, 1103; in Lycoming county, 1123; in Centre county, 1120; in Blair county, 1118; in southern Bedford county, 1114; Hyndman section of, 1117; Pulpit rocks of, 1109; on the Susquehanna, Chapter LXXVII, 1071; shales, 1048; on the Delaware and Lehigh, Chapter LXXVI, 1045; on Broadhead creek, 1054; fossils, Chapter LXXXI, 1125; fossils in Perry county, ...1089
Ormsby mine, Pittsburgh coal bed in, ...2546
Orr's mine, Sharon coal bed in,...1907
Orwigsburg anticlinal, No. V exposed upon, valley, ... 936
Osceola mine, Moshannon bed D in, ...2201
" Phillipsburg district, ...2200
Otto colliery, Southern coal field, ...2118,2121,2122
Oval valley, Lycoming county, "black marble" of No. IIc in, ...482,483
Overfield's large quarry, large flags of No. IX in, Wyoming county,..1593
Oxford type of rocks, N. J. (Syenite gneiss of Smock), described, .. 72

P.

Pacific slope, Sharon coal bed at, ...1909
Pack Saddle gap, ...1701
Packer No. 5 colliery, Western Middle coal field, ...2060
Paddy's valley axis, ... 871
Paint mines, Lehigh Water gap, ...1067,1176
" ore in Perry county, ...1184
Painted Post section, No. VIIIg, Chemung, N. Y., ...1379
Palaeozoic column, Pocono No. X sandstone in the, ...1629
" island, ...2590-1-5-8-9,2600-10-19-21
" river system, disappearance of. 48

Palisades, Hudson river trap, ...2621
Palmer coal bed, Southern coal field, ...2104,2115
" vein colliery, Southern coal field, ...2113-4-5
" tunnels, Southern coal field, ...2098,2101-2
Palmyra group of quarries, limestone No. II. 320
Panama conglomerate of New York, ...1519
Pancoast colliery, Northern coal field. 1968-72
" mine, bed D at, ...2305
Panther creek basin, Southern coal field, ...2086
Paradise furnace section, mountain limestone. 1799,1801
Parallel fault theory, New Red formation, ...2595
Park No. 2 colliery, Western Middle coal field, ...2077
Parker well, Columbia No. 4, No. VIIIq, Venango oil group, ...
" twp section, No. XII in N. Butler county, ...1897
Parkesburg artesian well in II, record of, ... 498
Parlor coal bed, Eastern Middle coal field, ...2040
Parnell's Knob, synclinal mountain of No. IV. 278-9-82,681
Parrish colliery, Northern coal field, ...1990-93
Paschall S. E. fossils, Norristown shales, ...2612
Passmore mine, Lower Freeport coal D at, ...2323
Path mountain, ...252,296.682
" valley, ...209,279,533
" " axis, ... 871
" " fault, ... 210
" " mines, limonite ore, ...252,304,357
Patrick's mine, bed B at, ...2353
Patterson section, No. VIIId, Tully limestone. 1321
" mines, Lower Productive group at, ...2348
" quarry, No. VI limestone in, ... 955
Patton limestone quarry, fossils of No. V in, ... 895
" bank, bed D, Jefferson county, ...2306
Paul mine, Pittsburgh bed, ...2491
Paultney slate belt, Vermont, described. 583
Paving stones of the New Red rocks, ...2628
Paxtang group of quarries, limestone No. II, Dauphin county,..... 322
Paxtonville, iron sandstone outcrop at. 815
" mines, Birds Eye fossil ore, ... 816
" quarries, block ore bed in, ... 817
" section, of No. V, referred to, ... 817
Payne's colliery, Southern coal field, ...2085
Peach Bottom, ...134-36-83,201
" " ridge, ...140-1
" " belt, slate series, ...130,203,555
" " quarries, roofing slate, ...555-81
" " series, phyllite beds of, ... 131

INDEX FINAL SUMMARY REPORT. 69

Peach mountain coal, bed, Southern coal field, ...2104-15-23
"   "   colliery, Southern coal field, ...2112
"   "   slope, Southern coal field, ...2102
Peaked mountain anticlinal, ...2082
Peat bog theory; formation of coal, ...1929
"   "   minerals, near Scranton, ...2021
Pebbly sandstone quarries, New Red near Yardleyville, ...2628
Peckville-Winton anticlinal, ...1960
Pen Argyle, slate of No. III, ... 584
Penlynn quarry, roofing slates of No. III, ... 605
Pennington range in Huntingdon county, limonite ores, ... 399
"   "   ridge anticlinal, ... 390
"   " Gas Coal Company, Pittsburgh coal bed, ...2510
"   " . ne, Pittsburgh coal bed, ...2510
"   " shaft No. 2, Pittsburgh coal bed, ...2512
Pennsylvania compressed rock waves, 422; zinc and lead ores, 437;
              why no Trenton oil and gas in, 494; underground geol-
              ogy, 26; Triassic deposit, 52; topography Middle dis-
              trict, Chapter LII, p. 681; Catskill in Western, 1573; the
              Corniferous in No. VIIIa, Chapter LXXXII. p. 1170;
              slates, 585; sea, 16; flagstone quarries, 569; formation,
              42; folded structure, 25; furnace ore banks, Chapter
              XXXII, p. 372; fossil ore outcrops, 752; gaps in moun-
              tains of Middle, 14; nomenclature of survey, 535; term-
              inal moraine of glacier, 51; Medino-Oneida mountain
              range, 709; limestone caverns, 429; anticlinal arches,
              687; model of corrugations, 704; No. IV, 678-9; No. I to
              VII—VIII and IX, 47; No. VI, Chapter LXVI, p. 917;
              No. VIIIa, 1228; No. VIIIb (Marcellus), Chapter
              LXXXV, p. 1203; No. VIIIc (Hamilton), Chapter
              LXXXVIII, p. 1247; No. VIIIf (Portage), 1354; nomen-
              clature of Hamilton, 1237; slates of Great Valley, 562;
              oil bearing counties, 1414; ore horizon, 414; No. VIII out-
              crops, 1038; palaeozoic formations, 898; Chemung, 1545-
              1434 to 1452; first survey, 809, 581; first survey, mem-
              bers of, 743; valley terraces. 50
"   Coal Company, 1981; No. 6 shaft, 1980; No. 1 shaft, 1970;
                No. 10 shaft, 1981; No. 11 shaft, 1980;
                No. 14 shaft, 1979-82; Dunmore. 1969-70
"   colliery, Western Middle coal field, ...2066
"   furnace ores, limonite; Centre county, ... 378
"   quarry, No. III roofing slate, ... 593
Penn's creek, ... 684
"   Narrows anticlinal, ... 368
"   valley, ...351,535,551
"   "   anticlinal, ... 368

Penny mine. Pittsburgh coal. 2533
Pentamerus limestone of No. VI, ...907
Pequea, ...133
" belt, ...109-10-17
" valley, ...270
Perkasie shales, New Red, 2608; fossils, 2609; coal, 2630; ores, 2632;
        trap, 2618-23; soil, ...2626
Perkiomen lead and copper mines, ...2631
Perry county section, 970; Centre Mills, 1088; Juniata gaps, 643; Montebello gap, 1271; Clarkes mill, 1013; Hamilton sandstone. 1273; fault, 1088-9, 1211-65; anticlinal, 1187; dykes, 455-6-8, 1459-63; No. V in, 746; No. VI in, Chapter LXX, 967; No. VI quarries in, 980; No. VII in, 1084 to 1090; No. VIIIa in, Corniferous, 1182 to 1184; No. VIIIb in, Marcellus, 1211; No. VIIIc in. Hamilton, 1266; Portage, VIIIf rocks in, 1363; No. IX Catskill in, 1601; Conglomerate in, 1557; Newport limestone in, 1594; Dillsville green sandstone, 1558; Kings mill sandstone, 1566; paint ore, 1184; Catskill iron ore; 1603; fossil plants, 1270; fossil plants, 1270; fossil fish spines, 198, 18, 29, 146. 294, 298; fossils. Hamilton, ...1304
" township, Berks county, flagstone quarries, ...569
Perrysville ridge, ...811-13
" anticlinal, ...1091,2292
" axis, ...2300-05,2423
Peter's creek mines. Pittsburgh coal and limestone. 2551
" (or Fourth) mountain, ...456-8-60 to 63,1569-72
Petriken opening. East Broad Top coal field. 2181
Petroleum Centre well record, ...1539
" of the Genesee, ...1334
Petrolia level, ...1542
" Evans well record, ...1543
Pettibone colliery, Northern coal field, ...1988 to 1996
Petty's quarry. No. VI limestone. 952
Pfoutz's valley, ...980,1185
Philadelphia belt, 45, 70-8,85,94, 103-7-18-19-21 to 25, 151-91. 1375; belt in Chester, Lancaster and York. Chapter XII. 127; system. 91, 203; gneisses. 90; rocks. newer gneiss. Chapter XI. 118; and Reading Coal and Iron Company,....1938,2144
Philipsburg-Houtzdale basin. Clearfield-Centre counties. 2134,2209
    -Osceola district. Clearfield-Centre counties. 2200
Philson (or Coleman) coal bed. Barren measures. 2418
" mine section. Barren measures. 2419

Phoenix colliery, Northern coal field, ...1978-80
" mine, Pittsburgh bed, ...2541
" Park collieries, Southern coal field, ...2122-23
Phoenixville and Warwick mines, magnetic ores of, ... 269
" mines, ores of New Red, ...2632
Phyllite belts of York and Lancaster counties, Chapter XIII. 133
Piedmont country, ... 178
" or Homewood sandstone (No. XII), ...1797,1891,2242
Pigeon Cove, ...1092
" " axis, ...1331
" " anticlinal, ...1092,1231
Pike county ....1570 to 1599
" " section, ...1411,1584
" " cascades, ...1327
" " Catskill, ...1585
" " geological map, ...1249
Pile mines, bed C', ...2261
Pincher (Fouse) tunnel section, fossil ore bed in. 830
Pine Brook colliery, 1969; creek basin, 1871, 2275; creek, synclinal,
1471; Forest colliery, 2105 to 2112; Grove quarry, 594; Grove
slate section, 1554; Grove narrows, 1185; Ridge anticlinal, 1116;
run mine, ...2548-50
Finedale colliery, Southern coal field, ...2163
Pinhook anticlinal, ...2547
Pipe ore pits, Nittany Valley limonites, ... 391
Piper bank, Lower Freeport coal D. 2342
" tunnel, fossil ore in; Shade Mt., ... 817
Pipher (Emory) quarry, No. III slate, ... 589
Pithole well, Venango group in. 1537-39
Pittsburgh region, description, ...2449
" " structure, ...2460
" " No. XV in, ...2538
" level, 1542; coal bed, 2449-57-65-69-71-75-82, 2503-17; little
coal bed, 2407; ore group section, 2434; limestone, 2457-63,
2558; oil wells, 1713, 1190; sandstone, ...2466
Pittston division, Anthracite coal district, ...1973
" gap, ...1827
" coal bed or "14 foot bed," Northern coal field, ...1980
Plastic support of formation, New Red, ...2600
Platt coal bed, Barren measures No. XIV, Somerset county, ...2420
" Franklin, ...498,649,1552,1616,1934,2008
Pleasant Valley, ...1971-5
Pleasantville, ...1537
" well record, ...1539
Plum Creek mine, Pittsburgh coal at. 2530
Plum Run ridge, ...1092

Plymouth colliery, Northern coal field, ...1994 to 97
Pocono mountains, 1372, 1651-3, 1811; mountains, escarpment, 718; mountains, plateau, 1568-71-89, 1947; knob, 1569; sandstone formation, 1681, 1783, 1837-59, 2292; rocks, 1663; sections in, 1635-7-9, 1641-3; sections, Conemaugh gap, 1703; sandstone synclinal, 1657; coals, 1631; coals, Hunter's cove, Perry county, 1655; coals, Tipton run, Blair county, 1677, 1679; Lewis tunnel, C. & O. R. R., ...1675-7
" formation No. X, Chapter CXI, 1629; on the Lehigh, Chapter CXII, 1635; inAllegheny mountain plateau, 1721; in Boyd's Hill well, Pittsburgh, 1713; in Cambria county, 1695; No. X, in Crawford county, 1749; in Fayette county, 1699; In Huntingdon county, 1659; in Sideling Hill, 1663, of Laurel Hill, 1709; in McKean county, 1729; Oil creek lake group, 1767; in Perry county, Chapter CXIII, 1655; in Somerset county, 1697; in Trough valley, 1789; at Turkey Nest, 1705; at Victor hollow, 1709; in Westmoreland county, 1699; in the Northern counties, 1717; in Southwestern Pennsylvania, ...1709
" fossils, ...1675,1689
Poe valley anticlinal, ... 309
Pohopoco mountain, ...1568
Point of Rocks, New Red district, ...2623
Point Pleasant trap, Gwynedd shales, New Red, ...2623
Pond creek, Marcellus formation along, ...1203 to 1209
" " buried valley; Marcellus formation in, ...1207
" " coal basin, Eastern Middle coal field, ...2025
Pool No. 1 mines, Monongahela river, 2546; No. 2 mines, 2548; No. 3 mines, 2550; No. 4 mines, 2554; No. 5 mines, 2557; No. 6 mines, 2560; Nos. 7 and 8 mines, ...2562
Poor House coal bed, Southern coal field, ...2104
Pope's Rock outcrop, Panama conglomerate, VIIIg, ...1529
Port Barnet fire clay, 1888; Blanchard, 1976; Jackson shark, 1019; Jervis, 917; Jervis section, 1013; Jervis quarries, 934; Jervis No. V at, ... 733
Portage formation No. VIIIf Chapters XCII, 1336; XCIII, ...1364
" deposits, conditions of, ...1348
" oil and gas, ...1352
" fossils, ...1343
Portland, Conn., fossils in Norristown shales, ...2614
Potato creek basin, vertical section of coal measures in, ...1878
Potomac, ...137 to 148, 271 to 273, 296
" valley, line of levels, ... 273
" water gap, Great Valley, ... 272
" exposures, same as on Juniata, ... 40

Potomac country, South mountain mass in, ... 185
" formation; division of the New Red, ... 2634
Potter county; geology of, ... 1434 to 36
" " No. XII in, ... 1872
" " No. XIII in, ... 2275
" " Oswayo synclinal, ... 1876
" " fish beds, ... 1462
.. " wells, oil borings, ... 1478-9
Pott run tunnel, coal exposures near, Western Middle coal field,..2055
Potts' slope, Southern coal field, ... 2099
Pottstown shales described, New Red sub-division, ... 2607
" " highest beds, ... 2608
" " fossils, ... 2608
" " soil, ... 2626
" " trap, ... 2618-23
Pottsville division, 2104; mountain, 1262; gap, 1723; coal basin, 89;
 coal basin synclinal, 126; tidal layers of red mud, ... 17
" conglomerate formation No. XII, Chapters CXVI, CXVII,
 1643, 1653, 1713-85, 1835-44-57-58, 2259-77; in the
 anthracite, 1853, 1920, 2075; limits, 1920;thick-
 ness, 1922; series, 2165; at St. Mary's, 1881-2;
 Knob synclinal, 1657; in Northern coal field,
 1949-53; in Eastern Middle coal field, 2025;
 in Western Middle coal field, 2056; in Broad
 Top basin, 1862; in Sullivan and Lycoming,
 1864; in Clinton, 1866; in Centre, 1867; in
 Cambria and Somerset, 1869; in Bradford
 and Tioga, 1871; in Potter, 1872; in McKean,
 1872; in Cameron, Elk and Forest, 1881; in
 Forest, 1885; in Jefferson, 1886; in Indiana,
 1889; in Westmoreland and Fayette, 1891; in
 Warren, 1892; in North Butler, 1896; in
 Crawford, 1898; in Mercer, 1902; in Lawrence, 1909; in Beaver, ... 1912
" colliery, Southern coal field, ... 2105-9-11 to 14
Poundstone opening, Sewickley coal at, ... 2523
Powder Mill coal bed, Northern coal field, ... 1977
" " anticlinal, separating Broad Mt. basins, Southern coal
 field, ... 2078
Powderly colliery, Southern coal field, ... 1963
Powel mines, thickness of fossil ore of No. V in, ... 845
" cove tunnel, thickness of fossil ore of No. V in, ... 845
Powell's quarries, fossils of No. VI in, ... 1028
" " section, No. VI limestone, Huntingdon county, .. 991
Powelton basin, Broad Top coal field, ... 2177-79
Powers and Brown mine, bed D at, ... 2305-8

5*

Poxono Island, Bossardville limestone exposures near, ... 930
"   " section, No. VI. limestone measures, ... 931
Precipitation of limonite in caves, ... 433
Prescott, level of divide, in Great Valley near Lebanon, ... 272
Presque Isle well at Erie, Corniferous formation VIIIa in, ... 27,1190
Preston No. 2 colliery, Western Middle coal field, ... 2061
Price coal bed, Barren measures No. XIV, ... 2415
" mine, section and analysis, Pittsburgh bed at, ... 2419
Priceville-Jermyn division, Northern coal field, ... 1958
Primrose coal bed in Eastern Middle field, 2030-40; Western Middle field, 2057-61-66-71; Southern field, 2101-12-22-31-37; Heckscherville basin, ... 2085
" mine, Pittsburgh bed, ... 2541
Prince's soapstone quarries, Montgomery county, ... 114
Prince Manufacturing Company's mines, paint ore of No. VIII,.... 1176
Problems of the survey of the New Red district, ... 2595
Productive coal measures, earlier beds, ... 1631
"   "   " No. XIII, Chapter CXXIV, ... 2155
Prospect colliery, Northern coal field, ... 1996-7
" shaft, anticlinal, Northern coal field, ... 1986
" Hill, Olean conglomerate, ... 1875
Pucketa creek coal section, ... 2530
Pulpit Rocks, Oriskany sandstone in Perry county, ... 1086
Punxsutawney basin, bed D in, the same as at Reynoldsville,.... 2306
Pyne colliery, Northern coal field, ... 1970-72
"   " Southern coal field, ... 2121-2-3

Q.

Quaker Hill, Bridgeport sandstone buried beneath, ... 761
"   " Marcellus iron ore in, Perry county, ... 1221
Quakertown coal bed, in Pottsville conglomerate, ... 1905-11
Quarries, marble of Chester valley, 1110; Erie county, 1512; **Lehigh** county, 604; Lower Harmony, N. J., 109; Peach Bottom roofing slates, 555; South mountains, York and Adams counties, 556; of No. VI in Perry county, 980; of **Wales and Vermont,** 556
Quarryville, hydro-mica slate, 133; trap dyke, ... 453
" railroad, limestone No. II section along, ... 486
Quebec group, discussed, 107, 299; fossils of the, ... 107,511-13
Queen's run, fish bed of No. VIII-IX; red conglomerate,............ 1466-7
"   " red beds of No. IX; exposure and sections, ... 1617-18-19
Quinn mine, Hublersburg Valley group, limonite ore, ... 378

R.

Rabbit Hole coal bed, Southern coal field, ...2116
Raby mine, magnetic ore in, Chester county, ...264
Raccoon ridge, Oriskany poorly exposed in, ...1085
Rainbow mine, Pittsburgh bed in, ...2520
Ralston-McIntyre basin rim of No. XI around, ...1835
    mines, bed E in, ...2187
    ore, in No. XI Mauch Chunk red shale, ...1841
Ramapo belt, containing serpentine everywhere, ...109
" mountain, trend of Archaean rocks in, ...63
Randall's section at Warren, oil sands of VIIIg, ...1492 to 1496
Rand sketch map, of steatite-soapstone outcrops, ...124
Rankin mine, Pittsburgh coal bed in, ...2538
Rathbone's ore bank, Upper Hamilton fossil ore in, ...1268
Rattling run gap coal beds in Southern coal field, ...2145
Raub Coal Company, red ash coal mined by; Northern coal field, ..1989
" ore in crystalline limestone, ...116
Rausch creek colliery, Southern coal field, ...2122-31
" gap mine workings, Southern coal field, ...2144
Ravine colliery, Northern coal field, ...1982
Rawley bank, bed D in, ...2324
" opening, bed D in, ...2333
Ray's Hill, Conglomerate No. XII series in, ...1863
" " synclinal, described; Broad Top field, ...2168
" mine, Pittsburgh coal bed in, ...2512
Raystown gap, fossil ore of No. V exposed in, ...850
Rea tract, Buck mountain bed, Western Middle coal field, ...2065
" bank, Washington coal bed in, ...2576
Reading hills, ...61,79,704
" and Durham hills, ...74
" levels, ...273,292
" quarries, limestone of No. II, ...357
Reakirt mine, bed D in, ...2324
Rector quarry, mountain limestone in, ...1849
Red ash coal bed, 1977-87, 2002; colliery, 1987-9; colliery near Minersville, Northern coal field, ...2113
" Bank Coal Company's mine, bed E in, 2349; mountain colliery, 2123; ridge, 885; ridge series, 897; rock, 1458-9; rock fish bed, 145-8-9; rock section, ...1443
Redstone coal bed, Up. Prod. Series,....2451-7,2463-9,2471-5,2503-17-33-53-62-64
" limestone, up. prod. series, ...2457-75,2517
" mine, Pittsburgh coal bed in, ...2519
" ridge group, Clinton rocks No. V, ...823-5
Reeder mine, Marcellus iron ore in, ...1220
Reevesdale tunnel, Southern coal field, ...2097-8,2100-2
Refuse in coal beds of Northern anthracite field, ...1952

Register pit, Pittsburgh coal bed in,	2558
Reiach, Lehigh metalic paint, No. VIII,	1176
Reliance colliery, Western Middle coal field,	2065
Remley's quarry, roofing slate No. III,	602
Reno colliery, Western Middle coal field,	2065
Renovo-Karthaus basin, description of,	2190-93
Reple's quarry, No. III roofing slate,	600
Report, First Survey,	2125-39
Repplier colliery, Southern coal field,	2085
Revenue colliery, Southern coal field,	2112
Revere boring, coal and oil,	**2630**
Rex's opening, Waynesburg coal in,	2561
Reynold's quarry, Venango group in,	1514
Reynoldsville basin, thickness of bed D in,	2306
" -Lisbon basin,	2292
" -Punsuatawney basin,	2293
" district,	2297
" synclinal,	2305
Rhaetic age of Gwynedd rocks, New Red,	2611
" name,	2589
Rhea mine, Pittsburgh coal bed in,	2470
Rhinesmith's quarry, No. VI limestone in Perry county,	981
Rhodes ore bank, Marcellus ore and Oriskany sandstone,	1227
Rice's small quarry, Corniferous,	1187
" Tudor &, quarries, Corniferous limestone in,	1186
Riceville shale, overlying first Venango oil sand,	1779
" " section,	1777
Richard bank, bed B in,	2214
Richardson colliery, Southern coal field,	2083-5
" farm well record,	1539
" mine, Waynesburg coal in,	2556
Richmond coal bed, Northern coal field,	1973
" drift and shaft, Northern coal field,	1973
Richmondale colliery, Northern coal field,	1955-7
Ridge shaft, Northern coal field,	1960
Ridgway section, red shale under conglomerate,	1735
" well, Chemung VIIIg rock section in,	1487
Ridley township, Leiper quarries, Granite,	97
Ringing Rocks trap, Black's Eddy, Bucks county,	2624
Risher mine, Pittsburgh bed in,	2546
Rittenhouse gap district, magnetic iron ores in,	269
Riverside colliery, Northern coal field,	1963
Road metal, New Red rocks,	2628
Roads of the New Red field,	**2592**
Roaring Run axis,	1890, 2347
Robb mine, Pittsburgh bed, in,	2471

INDEX FINAL SUMMARY REPORT. 77

Roberts' mine, Pittsburgh bed, in, ...2534
"   quarry, Peach Bottom roofing slate, ...138
"   run basin, Eastern Middle coal field, ...2034-37
Robertsdale colliery, Broad Top coal field, ...2177-81
"   slope, East Broad Top coal field, ...2183
"   No. XII at, ...1863
Robertson bank, Pittsburgh bed, in, ...2514
Robbins' & Jenkins' mine, Pittsburgh bed, in, ...2532
Robinson mine, bed B, in, ...2187
"   E., mine, bed A in, ...2314
"   quarry, No. III slate in, ... 594
Rochester mine, bed D, in, ...2297,2305
"   "   bed D in, ...2214
"   cannel mines, Alton upper coal, ...1877
Rochester and Pittsburg mines, bed D in, ...2297
Rock Cabin mines, bed E, ...2191
"   coal bed, Northern coal field, ...1965-72-98
"   exposures, Bucks and Montgomery counties. 2603
"   sample for comparison, Bucks and Montgomery, ...2603
Rocks of New Red field, ...2602
Rockford mottled limestone, bottom of Waverly group, Ohio, ...1787
Rockhill gap, Huntingdon county, ...631-53,713
Rockhill gap, Huntingdon county, 631-53, 713; section, ...645,779,799
"   mines, Huntingdon county, ... 817
"   slope, Huntingdon county, ...1230
"   trap, Montgomery county, ...2625
Rocktown tunnel, Southern coal field. ....2100 to 2102
Rocky mountains, ...18,46,54,192,198
"   "   clays of, ... 10
"   "   collections in, ... 193
"   ridge, Broad Top coal field, ...1174-84,1835-63
"   "   basin, Cook bed in. 2181
"   (or Red) ridge, ... 839
Roderick pit, Pittsburgh coal bed in. 2523
Rogers, H. D. ....1185 to 88-94,1421,1526,1838,2146
"   "   State map, ... 127
"   "   section and measurements, water gaps, ... 641
"   "   "   "   "   Danville, ... 748
"   "   "   "   "   Bellefonte gap, ... 647
"   "   "   "   "   Jack's Narrows, ... 645
"   "   "   "   "   Will's mountain gap, ... 649
"   W. B. gravels and cobblestones of Virginia, ... 189
Rohrabaugh's quarry section, No. VI limestone, ... 954
Rohrsville slope, Southern coal field, ...2085
Rome, N. Y. quarries, fossils found in, ...1044
Rondout quarries, hydraulic cement beds in, No. VI. 925

78   GEOLOGICAL SURVEY OF PENNSYLVANIA.

Roofing slate beds of No. III, Chapter XLVIII, ... 574
"   "   belt, ... 543
Rooker farm well record, ...1539
Rose (or Philson) bed, Barren measures No. XIV. 2263,2422
"   bank, Mercer lower coal in, 1905; bed D in, ... 2238
"   J. opening, Pittsburgh coal bed in, ...2510
"   crystal marble quarry, ... 109
Rosendale quarries, hydraulic cement bed in, ... 925
Ross coal bed, Northern coal field, ...1989,2004
"   mine, Pittsburgh coal bed in, ...2487
"   pit, bed D in, ...2306
"   ore deposit, limonite ore, ... 368
"   "   bank, Marcellus iron ore, ...1225-6
Rosston section, ...2158
Rostraver mine, Pittsburgh coal bed in. 2519
Rothwell, topographical map. Pittston division. Northern coal field,1974
Roulet (Hebron) anticlinal, ...1462-3-4-78
"   anticlinal valley, ...1435
"   fish bed, in Chemung-Catskill rocks, Potter county, ...1464
Round Knob basin, Broad Top coal field, ...2169-73
"   "   "   Kelly coal in, Broad Top, ...2170
Roundleys section, Clinton rocks, ...757-8,803
Rowe tunnel, Southern coal field. 2132
Royer's quarry, limestone No. II. 314
Royer & Dewees' shaft and tunnel, Marcellus iron ore in, ...1230
"   ridge, No. XII in. 1099
"   "   anticlinal, Marcellus and Oriskany on. 1229
Ruff bank, Pittsburgh coal bed in, ...2488
Rugh, J. mine, Pittsburgh coal bed in, ...2499
Rupert-Catawissa section of Hamilton No. VIIIc. 1269
Rush Brook gap, Northern coal field, ...1959
Russel bank, Pittsburgh coal bed in, ...2540
"   quarry, Salina in, 744; Helderberg fossils, ...1011,1508
"   section of Salina rocks, ...744-5
"   "   "   No. VI, ... 942
"   A. F. quarry, fossils of No. VI in, ...1025
Russian section of Tully limestone VIIIg, ...1311
Rutherford estate quarry, limestone No. II. 322
"   Beaver-Paxtang limestone belt, description of, ... 324
"   and Barclay slope, paint ores in, ...1176
"   tunnel mine, paint ores in, ...1177
Ryan's, Thos. quarry, description of, ... 598
"   section, VIIIg Chemung, ...1428

S.

Saddleback ridge, Hamilton No. VIII sandstone exposed in, ...1229
Salamanca conglomerate, description of, ...1530,1531,1532
Salem coal basins, Northern coal field, ...2003,2006
" " bed in Southern coal field, ...2116,2125
" " Company, coal in No. XII at, Northern coal field, ...2001
" colliery, Southern coal field, ...2116
Salina No. V, Chapters LV to LXIII, 721; of Lewistown valley, Chapter LVIII, 779; of Huntingdon valley, Chapter LIX, 819 in Bedford county, 839; in Bloomsburg red shale, 805, 823; upper limestone, 821; middle green shales, 822; upper limit of, 742; gypseous character of, 744; lower shales, 760, 773; Bridgeport sandstone, 761; variegated shales, 805, 764; fish beds, 764, 768; fossils, 775; upper gray shales, 776; fossils of, ... 889
Salisbury basin, ...1842,1869,2336 to 2343,2409,2450,2455,2461
" " analysis of Pittsburgh coal bed in,...................2465
" " of Somerset county, No. XV in, ...2461
" -Berlin basin, Barren Measures in, described, ...2416
" district, coals of the Lower Productive series in, ...2244
Salt-brine bearing formation (Pocono S. S., No. X), description, ...1629
" Lick twp.. section, general section of Umbral rocks in, ..1849
Saltillo fossil bed in Blair county, ... 853
Saltsburg basin of Indiana county, ...2336
" Coal Company's mine, Pittsburgh coal bed at,.............2507
" wells, Venango oil group in, ...1543
Sand vein fossil ore in Blair county, ... 853
Sand glass in Mapleton range, ...1103
" mines of No. VII, ...1094
" ridge, Oriskany sandstone in, ...1103,1005,1110,1111
" " anticlinal (Tadpole ridge), Centre county, described,...367,368
Sandrock coal bed, or Tunnel bed, Southern coal field, ...2116
Sandstone, gray, quarries, in New Red, 2627; for paving, ...2628
Sandy creek mines, Pittsburgh coal bed in, ...2529
" ridge, barite in, near Orbisonia, ... 448
" run colliery, Eastern Middle coal field, ...2029
Sarver pit, Pittsburgh coal bed in, ...2528
Savage mountain anticlinal. Hamilton rocks on, 1281; Chemung conglomerate, ...1551
Saw creek falls, section of Genesee, Tully and Hamilton rocks, ...1253
Saw Mill run mines in Pool No. 1, Pittsburgh coal bed, ...2545
Saxman's mine, Pittsburgh coal bed in, ...2485
Saxton section, Hamilton rocks in, 1281; Marcellus ore, ...1234
Saylor mine, Pittsburgh coal bed in, ...2462,2464
Saylorsburg iron ore diggings, ... 929

Scahonda tract, description of, Elk county, ... 1883
Scalp Level coal district, Somerset county, described, ... 2253
Scaur limestone of Great Britain and Belgium, ... 1789
Schaeffer mine, bed D at, ... 2255,2261
" J. mine, bed E at, ... 2256
" B. mine, bed C' at, ... 2254
Schafer mine, bed B at, ... 2237
Schmoele slope, Southern coal field, ... 2131
Schoharies region, prevalence of Dalmanites selenurus in, ... 1167
School House No. 5 opening, Pittsburgh coal bed in, ... 2540
Schooley colliery, Northern coal field, ... 1979,1980,1981,1932
Schroeder mines, analyses of bed B in, ... 2267,2269
" " section and analysis of bed B in, ... 2267
Schupstein mine, bed E in, ... 2256
Schuylkill county, No. VI in, Chapter LXVII, 936; vertical section of
    fault, 402; -Dauphin basin, description of, Southern coal
    field, 2138; -Dauphin basin, condition of coal beds in, 2141;
    and Delaware, series of New Red beds nearly identical,
    2603; region, contents of, anthracite coal, 2151; river (slate
    belt across, at Hamburg), No. III, 569; river, newer gneiss
    belt exposed along, 119; valley, gneiss, gorge, 120; water
    gap, 560, 631, 708, 709, 710, 726, 727; water gap, countour
    map of (see also page 666), 673; water gap, measurements
    of No. IV in, 643; water gap, No. V at, 733; water gap,
    No. IV at the, general description, ... 673
Scolithus linearis, Chapter XVII, ... 187
Scotch valley, Danville ore bed along Brush mountain in, Blair
    county, ... 354
Scott mine, Barnet bed in, 2177; bed C', 2262; Waynesburg coal, ... 2556
" Steel coal bed, in Southern coal field, ... 2083
Scranton division, Northern coal field, description of, ... 1966
Scrubgrass coal bed, in Armstrong county, 2358; Mercer county, 2390;
    absent in Beaver county, ... 2402
Shickshinny mountain terrace, Catskill rocks in, ... 1607
Seams' (Samuel) quarry, roofing slates of No. III, ... 594
Sears farm well record, section of Venango oil group, ... 1539
Seaton, T. mine, Pittsburgh coal bed in, ... 2472
Second (Coalmont) basin, Broad Top coal field, description of, ... 2171
" bituminous coal basin, No. XI in, near Karthaus, ... 1840
" " along the Susquehanna, coal measures of, ... 2203
" " coal basin, general description, ... 2232
" mountain, Catskill rocks in, Central Pennsylvania, ... 1569
" Upper Red Ash coal bed, Second Twin beds, Southern coal
    field, ... 2093
Sections, general definition, Chapter IV, ... 30
Sedimentary rocks of the field, New Red district, ... 2602

Segler mine, Pittsburgh coal bed in, ...2551
Seidel's quarry, No. VI limestone in, Montour county, ... 955
Selinsgrove anticlinal, description of, 1258; No. VI on, ... 937
"   "   No. VIIIa on the, ...1178,1257,1258,1262
"   No. VII at, ...1671
"   North and South dipping sections, Nos. V to VIII, ...958,961
Sellers, C. mine, Waynesburg coal bed in, ...2557
Seltzer's quarry, limestone of No. II in, Berks county, ... 312
Seneca colliery, Northern coal field, ...1982
Seneca Falls well, Medina red shales and sandstones in, ... 649
Seral conglomerates (see No. XII Pottsville conglomerate), ...1783
Seven Foot coal bed in the Broad mountain basins, ...2079
"   "   "   " in Northern coal field, ...1981,1997
"   "   "   " in Western Middle coal field, ...2056,2060,2065,2070
"   " knoll, Alton coal beds in, McKean county, ...1880
"   mountains, topographical features of, ...682,684
Sewickley coal bed,2451,2455,2466,2469,2471,2474,2493,2502,2516,2524,2533,2562,2564
"   limestone, ...2464,2470
"   mine, Pittsburgh coal bed in, ...2512
Seymour bed, Fall Brook Coal Company's mines, Blossburg basin,...2273
Shade gap, No. V rocks and ore beds in, ...779,808,817,818
"   mountain, No. VII in, near Peru, Tuscarora valley, ...1093
"   "   anticlinal, ...957,987,1077,1079
"   "   gap, No. IV in, ... 669
"   "   Granville gap, section of Danville ore beds in, ... 812
Shafer bank, bed D in, ...2303
Shafer's valley, group of No. IV mountains in, ... 682
Shaft coal bed, Northern coal field, ...1957
Shafton mine, Pittsburgh coal bed in, ...2512
Shamburg (well record), Venango oil group in, ...1530
Shamokin basin, synclinal, 1256; No. XI in, 1811; Trevorton division, ...2062,2066
Shaner mine, Waynesburg coal bed in, ...2556
Shank mine, bed D in, ...2246
Shape of the New Red of Bucks and Montgomery county fields, ...2501
Sharon coal bed, 1906, 1907, 1911, 1914; conglomerate, 1863, 1899; Garland-Olean conglomerate, 1785; group, 1860, 1897; (Marshburg) coal bed, 1863; -Olean conglomerate, general section of, 1749; upper shales and iron ore, 1906; well, sandstone of the Oil creek lake group in, ...1769
Sharp mountain, Nos. IV to XII in, 295; Mauch Chunk rocks in, ...1811
"   "   colliery, Southern coal field, ...2089,2141
"   "   gap below Tamaqua, Buck mountain coal in, Southern coal field, ...2097
Sharpsville sandstone, description of, in Crawford county, No. X,...1761

Shawangunk mines (lead) in No. IV rocks, ... 678
" mountain (Kittatinny mountain) No. IV in New Jersey, 676
Shawmut basin, description of, ...2283,2286
" and Meade run mines, Elk county, described, ...2287
Shearer tunnel, fossil ore bed of No. V in, ... 817
Shelly station trap, in New Red, ...2625
Shenango sandstone, ...1743,1751,1757
" shales, ...1749,1839
Shenk Brother's quarries, limestone of No. II in, ... 318
" and Herr's quarry, Lebanon city group limestone of II, ...315,316
Shephard bank, bed D in, ...2324
Sherman mine, Alton coal bed in, ...1881
" valley, ... 682
Sherwood's colored map, ...
" section, Fall creek conglomerate in, No. VIIIg, ...1556
Shickshinny mountain, No. IX in, 1570; Pocono No. X in, ...1651,1825,1829
" synclinal, Marcellus, Hamilton and Genesee rocks in,....1261
" mountain, White's section of, Nos. X and XI in, ...1608,1643
Shield's pit, Pittsburgh coal bed in, ...2500
Shinersville conglomerate, ...1864,2277
Shippen and Wetherill tract, coal beds in, S. C. F., ...2096,2099,2101
2096,2099,2101
Shohola bridge section, blue stone of No. IX. 1591
Short mountain, fault along, Canoe valley and Morrison cove, ...401,403
" " colliery, Southern coal field, ...2134,2135,2136,2137
" " gap, Danville ore beds poor in, ... 854
Shueys old quarry, Meadville lower limestone in, Crawford county,..1763
Shull, John, opening, bed D in, Jefferson county, ...2303
Shy Beaver section, Chemung shales in, Huntingdon county, ...1551
Sibley colliery, Northern coal field, ...1975,1989
Sideling Hill, Ashburner and Billin section in, Pocono S. S., 1663, 1665,
1667, 1669; Catskill section in. 1612; coal beds in tunnel
at, ...1671
Sides mine, bed D, Ligonier basin of Indiana county, ...2322
Siegersville limestone cove, limonite ore mines in,..................... 345
Silicious limestone, ...1791,1793,1795,1799,1842,1844,1846,1847
Silliman slope, Southern coal field, ...2097,2099,2101
Silurian, upper limit of, 899; upper fish beds, ... 773
Silver Brook basins, Eastern Middle coal field, 2046; collieries, ...2047
" creek coal basin, description of, in Elk county, ...2290
" " colliery, Southern coal field, ...2105,2108,2109,2110,2112
" " dam, shafting east of, in Southern coal field, ...2098
" " well, Pocono shales in, 1729; No. XI red shale in, ...1735
Sink Holes in No. II, Chapter XXXVI, ... 425
Sinking Spring valley, described, Blair county, ... 426
" valley zinc and lead mines in Blair county, ... 437,444

INDEX FINAL SUMMARY REPORT. 83

Six Foot coal bed, Northern coal field, ...1989
Sixth bituminous basin, limit of No. XI as red shale in, ...1837
Skidmore coal bed in the Broad mountain basins, 2079; in Heckscherville basin, 2083; the Southern field, 2097, 2109, 2121, 2130, 2136; the Western Middle field, ...2056,2060,2065,2070
Skinners Eddy section, No. VIII in, ...1446
Slate quarries, Northampton and Lehigh counties, Chapter XLIX, .. 588
Slateville, New, quarries, roofing slates, ... 580
Slatington quarries, No. III roofing slates, ...546,576,581
" region, ... 555
" section, character of roofing slate zone in, ... 576
Slattery's drift, Southern coal field, ...2102
Sleeman bank, Cook bed? in, Broad Top coal field, ...2181
Sligo Branch Coal Company's openings, bed B in Clarion county, ...2377
Sloane, T. mine, Pittsburgh coal in, ...2467
Slope coal bed, Grassy Island bed, Northern coal field, ...1957
" No. 4, or H, coal bed, Northern coal field, ...2006
Smicksburg basin, superiority of bed D in, Jefferson county, ...2299
Smith Grove section, block ore series of No. V in, ... 817
" bed C', 2325; Pittsburgh coal in, ...2467,2470,2472
" William, bank, bed D in, Indiana county, ...2342
Smithfield, LeBarr's quarry in, Bossardville limestone, ... 924
Smethport, ...1485
" section and well, ...1488
" well No. 1, ...1482,1485,1487,1488
Snake Island coal bed, "Rock bed," Northern coal field, ...1998
Snodgrass quarry, ...1765,1901
Snowden quarry, slate of No. III in, ... 589
Snow Hill mine, Pittsburgh coal bed, ...2521
Snow Shoe coal basin, description of, ...2196
Snyder Coal Company's shaft, Mercer lower limestone in, ...1901
" county, No. V in Union and Northumberland counties, ... 870
" " No. VI in, Chapter LXXI, ...983 to 987
" " No. VII in, ...1071,1077
" " No. VIIIa in, ...1178
" " No. VIIIc in, ...1276
" quarry. black beds in limestone of No. VI, ... 954
" ridge. upper conglomerate of No. VIIIg in, ...1551
Soil of the New Red, character. 2626
Solebury mountain trap, ...2619,2622,2623,2625
Solomon's gap, No. XI red shale in, ...1819,1825
Somerset county, Mauch Chunk No. XI rocks in, described, ...1842
" " No. XII in, ...1865,1870
" " No. XIII in, ...2241
" " No. XIV in, ...2114

Somerset Wilmore (First) basin of, ...2247
Sommersville bank, bed C, Clarion county, ...2376
Sonman shaft, record of, Cambria county, bed C' in, bore holes, 2222,2223
Sopper pit, Pittsburgh coal, ...2524
Southern anthracite field, 1919; anthracite field in Dauphin county, 1653; coal field, condition of the coal beds in, 2076; coal field, contents of, 2150; coal field, general description of, 2072; coal field, Pottsville conglomerate in, 1854; mountain, Blue ridge-Highlands range, described, ... 291
South mountain, geology of the, Chapter XIV, 142 to 151; mines along the foot of, 234; Blue Ridge range, relation to No. II uplift, 291; rock series described, 144; of Cumberland and Fayette counties, Dr. Frazer's sections of, 45; of Cumberland and York, Fayette and Adams counties, 62, 70; of Eastern Pennsylvania, 40; foot, 238; end of, 238; end spurs, 239; west face, 248; uplift, ...291
" Pyne colliery, Southern coal field, ...2120
" Salem coal bed, Southern coal field, ...2116,2125
" Valley Hill, geology of, Chapter XIII, 133 to 141, 176; Potsdam S. S. in, 175, 184; limonite ore in, ...211,452
" Valley Hill trap dykes, ... 452
" " " slate belt, Chapter XIX, ...199 to 204
" Wilkes-Barre collieries, Northern coal field, ...1991 to 1998
Spangler mine, beds C and C' in, ...2240,2246
" drift, Southern coal field, ...2125
Specific gravity of anthracite, ...1928
Specht mine, bed B, Somerset county, ...2231
Speer coal bed, not workable in Huntingdon county, Broad Top coal field, ...2175
Spencers slope, Southern coal field, ...2115
Splint mine, Alton Lower coal bed, McKean county, ...1877
Spohn coal bed, Southern coal field, ...2115
Sponge coral reefs of No. VI, Chapter LXX, ...1020
Sprague mine, bed D, Reynoldsville district, Jefferson co.,..2297,2305,2306
Spragueville section, Starucca S. S. group in, ...1587
Spring mine, Alton upper coal, ...1877
Spring Brook colliery, Northern coal field, ...1978
" Hill mine, Pittsburgh coal, ...2528,2531
" mountain basin and Silver Brook basin, described, Eastern Middle coal field. 2046
Springfield mines, ...2431
" township section, No. XI formation, ...1849
Spruce Creek anticlinal described, ... 367
" " gap, No. IV at, ...655,657
" " tunnel, No. II in, 401; No. VII in, ... 631
Squaw Hollow anticlinal, Moore's ore bank near, ...1225

St. Clair shaft, Southern coal field, 2110; bed E at, Indiana county,...2331
St. Louis limestone described, ...1791
St. Mary's basin, part of Fourth bituminous basin, described, ...2283
"   mines, bed B at, ...2280
"   Toby creek, or Dagus coal basin, ...1876
Standard mine, bed D in, ...2298
"   " section, Pittsburgh coal in Connellsville basin, ...2489
Standing Stone mountain, Portage rocks in, ...1356
Stafford colliery, Northern coal field, ...1969,1970
Stair tunnel, fossil ore bed in, ... 817
Stanton coal bed, "Five Foot bed" in, Northern coal field, ...1995
"   colliery, Northern coal field, ...1987,1992,1994,1996,1997
Star slate quarry, No. III slates, Lehigh and Northampton, ...590,606
Stark colliery, Northern coal field, ...1978
Starr slope, Southern coal field, ...2113
Stephens and Butler mine, Pittsburgh coal bed in, ...2519
Sterling No. 3 mine, bed D at, Centre county, 2201; No. 8, 9, 11 and 13;
    bed D at, Cambria county, ...2240
Stevens' colliery, Northern coal field, ...1979
Stevenson's section at Saxton, Chemung-Catskill rocks and fos-
    sils, ...1562,1563
"   Hindman section, Bedford county, Oriskany S. S., ...1135
Stiger, Adam opening, Pittsburgh coal bed at, ...2508
Stillwell's ridge, outcrop of Oriskany S. S., Fulton county, ...1092
Stineman's mines, bed B at, ...2227
"Stiper Stones," of Wales; equivalent to Chicques sandstone,.......... 167
Stiver bank, bed D at, ... 342
Stoler tunnel section, fossil ore bed of No. V in, ... 830
Stone mine, Pittsburgh coal bed at, ...2548,2549
" mountain, Oriskany vertical in, ...1107
Stony Brook beds, White's, top of VIIIg, described, ...1564
" creek basin, Eastern Middle coal field, ...2034
" Hill mine, Pittsburgh coal bed at, ...2521
" ridge, iron ore bogs along base of, Corniferous rocks, ...1175
Stormville section, conglomerate, 923, 924; Oriskany S. S., ...1058
Stout mine, Pittsburgh coal bed at, ...2509
Stover (ore deposit), on Brush Valley anticlinal, ... 368
Straighthoof bank, bed D at, ...2303
Street's run mine, Pittsburgh coal bed at, ...2546
Strike curves in the New Red district. 2598
Stroh colliery, Southern coal field, ...2125
Stroudsburg anticlinal and synclinal, outcrops of Hamilton in, ...1254
Strouse mine, bed D at, ...2306
Structure (geological), of New Red, 2600; compared with Connecticut
    valley, 2600; of Eastern Middle coal field, 2023; of Northern
    anthracite coal field, 1947, 1954, 1960, 1968, 1976, 1985, 2001;

of Southern coal field, 2086, 2095, 2106, 2118, 2127, 2133, 2139; of Western Middle field, ...2050,2053,2059,2063,2067
Stuchart and Craven opening, bed E in Indiana county, ...2338
Stump run, coal beds on, Southern coal field, ...2132
Sturgis opening, Pittsburgh coal. 2497
" shaft, Northern coal field, ...1964
Sub-division of the New Red by Rogers, of the formation, ...2606,2607
Sub-Olean conglomerate of McKean county, ...1645
Sugar Notch collieries, Northern coal field, ...1989,1994,1997
Sullivan county, No. XII in, 1864; No. XIII in, ...2184
" pit, Pittsburgh coal near Fairview, ...2563
Summit Hill quarry, South mountain bed, Southern coal field, ...2089
" slope, Southern coal field, ...2137
Sumneytown copper, New Red formation, ...2632
Sump coal bed, Franklin colliery, Northern coal field, ...1994
Sunbury-Catawissa basin, ...1262
Susquehanna Coal Company's Nanticoke colliery, Northern coal field,2004
" county, Chemung in, ...1439 to 1443
" gap, No. V in, 726; iron S. S. and block ore, 755, No. XI
in, ...1825
" gap above Harrisburg, conditions of No. IV in, ... 637
" river district, coal measures of Cambria county, 2238; limestone quarries on the right bank, Chapter XXVII, 324; No. VI on the, Chapter LXVIII, 939; No. VI on the, Chapter LXIX, 957; No. VII on the, Chapter LXXVII, 1071; (North Branch) No. VIIIa on the, 1180; No. VIIIc Hamilton on the main, 1255; No. IX Catskill on the, and Juniata, 1601; (North Branch) No. IX Catskill on the, 1604; (North Branch) Portage on the, 1364; No. II limestone quarries on the, ... 323
" limestone quarries west of the, ... 324
" river section of newer gneiss belt of York county... 128
" valley, character of the mountains of No. IV in, ... 686
" river gap, No. IV at the, 669; No. V at the, 735; section at Rockport, thickness of Chemung, Portage and Genesee rocks, ...1557
Suter bank, Glenwood Coal Company, bed D in, ...2323
Swamp, the, of Nockamixon township trap, New Red, ...2618,2625
Swank's mine, bed C' in, Somerset county, ...2250
Swartz's ridge, Selinsgrove upper S. S. in, ...1255
Swatara Falls colliery, Southern coal field, ...2121
" quarries, limestone of No. II, ...319
Swift Creek colliery, Southern coal field, ...2101
Switzer's section of Chemung flags, Erie county, ...1500
Sykes and Jones bank, Kittanning and Clarion coal groups in, ...2214
Syracuse salt deposits, ...889

T.

Taconic system of rocks, discussed, ... 552
Tadpole ridge (Sand ridge) anticlinal axis, described, ... 367
Taggart's mine, Washington coal bed section in, ... 2575
Tallman's mine, Pittsburgh coal in, ... 2471
Tamaqua (Middleport) division, Southern coal field, described, ... 2094
Tangascootac coal basin, ... 2191
Tannerdale mine, Dagus coal B in, sections, ... 2280,2284
Tanner's Hill quarry beds, No. VIIIg of Warren county, ... 1496
Tarr farm (well record), Venango oil group, ... 1539
Tatesville tunnel, fossil ore of No. V in, ... 845
Taylor bank, East Broad Top coal field, analysis, ... 2182
" colliery, Northern coal field, ... 1970
" quarry section, mountain limestone Nos. X-XI, ... 1799
" slope, Southern coal field, ... 2115
Taylortown coal bed, "Waynesburg B coal," described, ... 2578
" limestone Ib., Washington county group of No. XVI, ...2577
Taylorsville colliery, Southern coal field, ... 2085
Terminal moraine, ... 2015
Terrace mountain, topographical features of No. IV mountains in, 681, 697; Catskill rocks in, 1574, 1609; coal in, 1659; Pocono sandstone of, 1612, 1663; Riddlesburg gap section in, 1661, 1663; Shoup's run gap section in, ... 1659,1661
Tettemer's copper, New Red, ... 2632
Thayer's gas well, ... 1797
Thickness of the New Red formation and its sub-divisions, 2607; old estimates from dips, 2594, 2602, 2603; old ideas, 2594, 2595; uncertainty, ... 2602
" " gaps between exposed beds computed, ... 2604,2605
" " No. III, Chapter XLVI, ... 557
Third bituminous basin, in Cameron county, described, ... 2289
" coal bed, Northern coal field, ... 1957,1980
" mountain sand of Oil creek lake group, ... 1771
" (Powelton) basin, Broad Top coal field, ... 2174
" Upper Red ash coal bed, Southern coal field, ... 2093
" (Westover) basin, of Cambria county, ... 2238
Thomas Iron Company's mine, limonite near top of No. II limestone, ... 346,347
Thomas Iron Company's mines, Lehigh county limonite, ... 233,234
" " " " limonite along Mountain creek, ... 242
" " old pit and banks in hydromica belt south of York, ... 222,223
" mine, bed E in, Somerset county, ... 2259

Thomaston colliery, Southern coal field, ...2083,2084,2085
Thompson mine, Pittsburgh coal section, ...2541
"   opening, bed D in, ...2332
Thorn creek, Marshall No. 21 well, Venango group in, ...1543
Three Springs fault, ...1228
"   " Huntingdon county, fault of, ...1687
"   " (section) No. VI limestone shales in, ...1027
Tide plain of the Southern States, aerial geology of, ... 17
Timber ridge, conglomerate upper Chemung in, ...1552
Time, geological, Chapter II, ... 16
Tioga colliery, Southern coal field, ...2101
" county, Scolithus in Portage rocks in, 190; Chemung bluestone
quarries, ...1593
"   " fish beds, ...1458
"   " No. XII in, ...1871
"   " No. XIII in, ...2263
" river section, Nos. IX to XII, ...1584
Tipton run coal analysis, 1683; coals (No. X), 1643, 1679; fossil plants,
description of, 1689, 1691, 1693, 1695; mines, Blair county,
1631,1681,1683,1685,1687
Titman, shaft ore, red hematite in, New Jersey, ... 116
Titus bank, Pittsburgh coal in, ...2563
Toby basin, No. XIII coal measures in, Elk county, ...2283
" creek basin, Elk county, described, ...2279
Tocquan axis, newer gneiss upon in York county, ... 129
" belt, anticlinal belt of gneiss, ...134,135,136
Tollgate pipe ore range, thickness of pipe ore deposits in, ... 387
Tomhicken colliery, Eastern Middle coal field, ...2035
Tomlinson mine, Saw Mill run, Pittsburgh coal in, ...2545
Tompkins colliery, Northern coal field, ...1979,1981
Tonolowa ridge creek, No. VII in, ...1092
Top coal bed, "Shaft coal," Northern coal field, ...1957
Topographical features of Middle Pennsylvania, Chapter LII, ... 681
Topography of the New Red regarded. 2597
Towanda anticlinal, described, ...1436.1443
"   " axis, Chemung shales of Bradford county on, ..1444
" basin, absence of fossils in Chemung of Bradford county,...1419
" mountain, Catskill rocks on, ...1437
Towler and Hunt well, No. 3, No. XII Olean conglomerate in, ...1875
Town's quarry, fossils (gyroceros trivolve) plentiful at, ...1168
Tracy coal bed in Southern coal field, ...2103,2114,2123
"   "   " " Western Middle coal field, ...2057,2062,2066,2071
Transition beds IX and X, ...1643
" beds between Nos. XI and XII, ...1921
" layers (sub-Pocono), ...1647

Trap, age, in the New Red formations of Bucks and Montgomery counties, 2623; areas described, 2618; conformable sheets, overflows not dikes, 2621; deposits, extent, 2616; dikes in the field, 2622; dikes in No. II, Chapter XXXVIII, 451; for building stone, 2628; in place, rare, 2618; lithology, 2623; overflows in the field, 2621; paving stones, 2628; soil, ...2626
Trenton Cut-off trap dyke, described, ...2617,2620,2623
"   oil and gas in Pennsylvania, absent from, ... 494
Tremont mine, Pittsburgh coal, ...2519
Trevorton colliery, Western Middle coal field, ...2067
Tripp colliery, Northern coal field, ...1970,1972
Triumph well, thickness of Venango Third sand, ...1491
Trixlertown copperas mine, bore hole records in, ... 350
Trout's and Baker's iron ore mines, ...1000
Trough creek church, section below, mountain limestone X-XI, ...1799
"   "   section, Lackawaxen conglomerate, ...1551
Troxall's quarry, Magnesian limestone beds of No. II, ... 329
Troxell's quarry, analyses of limestone beds, 303; Bossardville beds of No. VI, ... 955
Troytown mine, Pittsburgh coal in, ...2521
True Blue slate quarry, No. III slate, ... 592
Tucker's colliery, Southern coal field, ...2102
Tudow and Rice's quarries, limestone, Corniferous, ...1186
Tully limestone No. VIIId, Chapter XC, 1307; area, 1313; fossils, 1312;
    horizon, 1315; in Pennsylvania, ..1317 to 1322
Tunnel coal bed "Sand Rock" bed, Southern coal field, ...2116,2125
"'   quarry, No. III roofing slate, ... 604
Turkey ridge, ...980,990,1079,1085,1091,1185,1213,1269
Turner mine, Pittsburgh coal, ...2467
Turn's steep bluff and terrace, Bossardville limestone in, ... 930
Tuscarora creek, quarry on, plant fragments in Catskill flagstones, 1582; mountain, Clinton rocks in, 753; fossil ore poor on east slope, 818, 1236; valley, Corniferous and Oriskany, 1222; valley gap, south outcrop of No. VI limestone in, synclinal, ... 990
Tussey mountain, No. II and III along, 363; limonite ores from, 379, 381; cross section through Springfield mines from Lock Haven to, 404; Medina sandstone of No. IV, through, 655, 661; Oneida No. IV feature of, 657; description of, 684; ravine system of No. IV in, 698; fossils of Oneida and Medina No. V in, 715; Salina upper limestone in, 821; Saltillo ore bed along, 824, 825; Barree lower shales in, 826, 827; sections of fossil ore bed of No. V along foot of slope of, 830, 844; Clinton Middle shales along, ... 836

Tussey mountain, estimate of No. IV in, ... 631
" " Southern range, analyses of fossil ore beds in, ... 844
Tuttle creek, section of, Olean formation, ...1894
Twin coal bed, Northern coal field, 2004; Southern coal field, 2109;
    Broad Top coal field, ...2173
" colliery, Northern coal field, ...1978
Tyrone, Cemetery ridge, ores of No. VI in, ...1002
" City gap, block or keel ore bed., ... 857
" gap, measurement of No. IV in, ...657,659

U.

Udree ore bank, limonite ore and analysis of, Berks county, ... 236
Ulysses (see Homer) anticlinal valley, Chemung in, ...1435
Umber, in Buckingham mountain, deposit of (New Red), ...2633
Umbral formation of H. D. Rogers survey, 1858 (No. XI rocks), ...1805
" limestone (mountain limestone), ...1793
" ores, absent from eastern slope Allegheny mountain in Som-
    erset county, ...1842
" ores in Fayette county, ...1851
" red shale (see Mauch Chunk red shale), ...1783
" rocks section on Youghiogheny river (Stevenson), ...1845
Umstead's cross roads, lime flagstones of Oriskany at, ...1076
Unadilla, forks of the, quarry containing oldest land plants, ...1244
Underwood slope (see Mumper mine) magnetite, 258, 259, 260; new
    mine, magnetite, ... 258
Union bridge, marble in the Sericite belt of Maryland, 136; canal, con-
    tact of No. II and No. III along, 288, 289; city, Third sand of
    Venango group at, 1506, 1508; county, west bank of the river,
    Lewisburg Company's limestone quarries, 954; county, No. VI
    in, Chapter LXXI, 983, to 985; county No. VII, in 1071, 1077;
    Deposit Furnace Company's quarry, limestone of No. II, 320;
    furnace, caverns in No. II. limestone near, 427; furnace, Bos-
    sardville limestone beds of No. VI at. 985, Springs, plaster beds
    of No. VI at, 912; Springs, Gypsum quarries at, 914; township,
    Centre county. Alexander's limestone No. VI in, 1002; town-
    ship, Centre county, Portage flags in, 1355; township, Tioga
    county, upper Chemung strata well exposed in, 1444; Snyder
    and Northumberland counties, No. V in, Chapter LXIII, ... 879
Uniontown coal bed, 2451, 2452, 2455, 2474, 2502, 2553, 2562; coal and
    limestone, 2537; limestone, 2455, 2464, 2516; sandstone, ...2454
Unionville. No. VI limestone used for furnace slag at, 1002; sponge
    corals of No. VI at quarry near, 1030; Marcellus iron ore
    and shale below, 1235, 1282; Oriskany at, ...1356
United mines, Nos. 1 and 3, Pittsburgh coal bed in, ...2488
Unityville, Chemung layers near, ...1467

INDEX FINAL SUMMARY REPORT. 91

Upper Barren Measures series, Chapter CXXVIII, 2565; Juniata river, Caudagalli fossils found in, 1043; Juniata river, No. VII on the, 1103; Juniata waters, character of Oriskany S. S. on, 1037; Lehigh coal basin, 2025; Productive or Monongahela river series No. XV, Chapter CXXVIII, ... 2448
Upton's quarry, exposure of Panama conglomerate in, ... 1522
Urban P. O. also axis, No. V rocks on. 963
Urey mine, section of bed D at, ... 2322
Urich's, S. quarry, No. II limestone in, ... 315
Utilization of small sizes of anthracite coal, ... 1933

V.

Vail station fault, ... 869
Valley mine, section of Pittsburgh coal bed at, ... 2491
VanArtsdalen quarry, crystalline limestone in, ... 80,102
VanArsdale's quarry, gypsum crystals in, 450; grapite in, ... 478
" limestone quarries, apatite cystals found in, ... 114
VanAuken's quarry, Decker's ferry sandstone in, ... 930
" " analyses of quarry beds. 932
Vandergrift mine, Pittsburgh coal bed at, ... 2558
Vanderslice's old slate quarry, Hamilton No. VIII, ... 1265
Vanuxem's analysis (plumbago), Bustleton quarry. Bucks county,... 478
Van Voorhees' pit, Pittsburgh coal bed at, ... 2563
Venango-Butler oil belt, Chemung formation in, ... 1372
" -Clarion oil fields, ... 1544
" county, 1405, 1414, 1415, 1430, 1438, 1866; diminished interval between First and Third sands, 1405; Venango oil group in, 1415; fossiliferous limestone in, 1430; Chemung conglomerate in, 1438; No. XII in. 1866, 1895; oil belt, 1430; oil sands, 1489; Venango group in. ..1537
" first oil sand, ... 1779
  oil field. 1484
  oil region. 1725
" oil sands, Chapters CII, CIII, 1497 to 1543, ... 1747
Venetia mine, Pittsburgh coal bed at. 2551
Venture mine, Pittsburgh coal bed at, ... 2545
Vermont black marbles, fossil life of, 513; Brandon iron mine, lignite, 249; Green mountain rocks, 45, 61; Highland range in, 159; sections across, 161; geological section crossing, 58; slate belt of Fairhaven and Paultney, 583; slate quarries, 556; slates, ... 585
Vernon mines, conglomerate No. XII at. 1851
Verona axis, bed F, shafted upon, Allegheny county, ... 2405
Vespertine sandstone (see Pocono S. S.). 1733
Viaduct anticlinal, description of, in Somerset county, ... 2241
" axis. 1841,1842,1869,2219,2227,2256,2412

92   GEOLOGICAL SURVEY OF PENNSYLVANIA.

Virginia. Blue Ridge range of, 40, 142, 1375; geology of, same as South mountains of Pennsylvania, 24, 40; geology of West, same as Western Pennsylvania, 40; Low Moor iron ore of, 1220; New Red, 2611, 2613, 2615, 2616, 2629; two ranges of limonite iron ores, 253; roofing slate of No. III, 585; Wythe county zinc ores, 437; and West Virginia, Flat Top coal field of...1857
Von Storch colliery, Northern coal field, ...1972

W.

Wabash slope, Southern coal field, ...2101
Waddell's new colliery, Northern coal field, ...1962
Waddle's bank, bed E in, ...2337
Wadesville shaft, Southern coal field, ...2110
Wadsworth, the azoic system, ... 59
Waffington mine, bed D in, ...2230
Wagoner's Mill section of Clinton upper lime shale, ... 759
Walbeck mine, bed D in, ...2322
Walcott well, thickness of Devonian and Silurian rocks in, ... 649
Waldron's ridge, thickness of medina sandstone in, ... 649
Walker (Hiram) mine, bed C in, ...2247
" (William) mine, Platt bed in, ...2420
Walkill valley, No. IV in, New York. 677
Wallace and Rothermel old slope, Southern coal field, ...2115
Walpack bend, cross section, Oriskany sandstone No. VII, ...1049
" ridge, ...1049,1172,1203-6-7
Walston mines bed D in, ...2301
Walter's mine, fossil ore of No. V in; fossils in, ... 896
" bank, Pittsburgh coal in, ...2524
Walton mines, Pittsburgh coal in, ...2546-48-49
" brownstone quarries, Trias formation, ... 320
Wanamie colliery, No. 18, Northern coal field, ...2003-4
" " No. 19, Northern coal field, ...2003-4-5
Wapwallopen valley, No. XI in, ...1827
" anticlinal, Montour's ridge, ...1819
" section, Genesee slate beds and shales, ...1329
Ward's ridge, synclinal outcrop Chemung VIIIg rocks, ...1551
Warfordsburg anticlinical (Pigeon Cove) No. VII exposed on, ...1092
Warner opening, Waynesburg coal in, ...2555
Warren sections, Chemung rocks, ...1492-3
" county, 1489, 1523; oil field, 1372; oil horizons, 1491; oil sands, 1486-89 to 96, Chapter Cl, 1489. Hosmer conglomerate, 1508; No. XII in, ...1893
Warrior's Mark, Lovetown range, Nittany Valley limonite ores,... 391
" ridge, ...819 to 22,1039-93,1103-7 to 16,1232 to 34
" Run, cross section, four ore horizons on, ... 387
" " colliery, Northern coal field, ...2003-4

INDEX FINAL SUMMARY REPORT. 93

Warsaw group, in Missouri, thickness of, ...1791
Warwick mines, serpentine geodes in gneiss, ... 103
" and Phoenixville mines; magnetic iron ore, ...263-269
" (Jones) mine, limonite in Primal slate, 205; magnetic
 ore at, ...256-7-67
Wasatch mountain section, Olenellus and Pre-Cambrian rocks, ... 192
Wash. deep deposit of, near Kingston, Northern coal field, ...1982
Washington anticlinal, 2529; coal bed, 2092, 2575; little coal bed, 2577;
 lower limestone No. II, 2568-74; middle limestone, 2573;
 upper limestone, 2570; colliery, 2305; sandstone, ...2577
 county, 1073; group, 2566-9; No. XIV in, 2446; No. XV in,
 2524; gas wells, ...1797
" -Butler, Allegheny region, Venango oil group, ...1544
" slate quarry, No. III roofing slate, ... 584
" quarry, roofing slate No. III, ...591,605
" -West quarry, slates of No. III in. ..,................. 580
" township, Northampton county slates, quarries, ... 590
" " Lehigh county slate quarries, ... 604
Water, deep wells, localities, New Red district,.....................2634
" gap, 928, 729; of Middle Pennsylvania, 708; of Delaware, 271;
 Chance's contour map, ... 575
 lime, Salina, described, ... 804
Waterloo gap, Oriskany sandstone in, ...1091
" quarry, Corniferous limestone exposure; fossils, ...1163-7
Watson ore deposit, Centre county limonite ore, ... 368
Watsontown anticlinal (Milton axis), Catskill rocks on, ...1607
Watts analyses, limonite in York county, phyllite belt, ...220-21
Waverly formation, Ohio (Pocono sandstone group),.................1711
" group, derivation of name; description of, ...1785-87
" rocks and fossils, description of, ...1779
Wayne county, ...1046,1553 to 1587
" " New York Clyde well, Medina formation in, ... 649
" " Oil Company, boring through Catskill rocks, ...1587
" station, Kennedy's granite quarry, ... 97
Waynesburg coal bed, No. XIV, 2448-51-2-74-89,2501-15-27-38,2553-6-9-60-3-8
 " sandstone, base of Washington county group, ...2579
 " limestone, ...2454-74,2516
 " axis, ...2294,2301-10
 " synclinal, Ocean mines, Pittsburgh district, ...2535
Weaver's run section, Marcellus iron ore absent from, ...1233
" tunnel section, fossil ore, ... 830
Webster colliery No. 3, bed B in, ...2226
Weehawken, New Jersey, fish fossil, ...2614
Weimer's gap, Oriskany rocks in, dip of, ...1113
Weirick's well, first oil sand (upper gas sand) in, ...1414
Weisenburg township, Berks county, quarry, slate of No. III,...... 611

Weissport anticlinal, Marcellus upper strata on, ...1206
Welchtown quarry, No. III slate in Lehigh county, ...604
Wellersburg, or Southampton, coal basin, Somerset county,..1871,2241,2414
Wells at Northampton, Mass., deep boring in New Red, ...2602
Wells' bank, bed C' in, ...2325
Wellsborough anticlinal, ...1436-59-71
" belt, Chemung and Catskill rocks....................1436-7
Welsh mountain, 78, 101-3-73, 453; region, 70-8-9, 80-2-7, 94, 166-73; azoic of North Chester county, 75-6; anticlinal, 761; banks, ... 228
Wenona mine, Pittsburgh coal bed in. 2537
West Branch valley, 867; Conshohocken furnace quarries, 81; Duncannon dyke, 458-60-3; End collieries, 2003-4-6; Eureka collieries, 2300-1; Hickory well, 1491; Lebanon-Salisbury basin, 2465; Lehigh colliery, 2100-1; Marlboro (Barnard's) quarry, 114; Shade mountain, 683, 779, 1091-2, 1222; Virginia beds perhaps coeval with Norristown shales, 2590, 2615; Washington quarry, ...580-94
Western Middle anthracite field described, 2049; contents of, 2149;
. Pottsville conglomerate, ...1854
" Northern anthracite field, ...2007
" Pennsylvania beds perhaps coeval with Norristown shales, 2590,2615
Westinghouse well at Pittsburgh, 4,685 feet deep......................26.27
Westmoreland county, ores of No. XI in, ...1849
" and Fayette counties, 1789, 1833-42; No. XI in, 1843;
No. XII in, 1891; No. XIII in,2362
" and Fayette counties W. of Chestnut ridge, No.XIV,2432
" and Fayette, No. XV, coal measures, ...2472
" mine, Pittsburgh coal bed in, ...2511
" shaft, Pittsburgh coal bed in, ...2506
Westover mine, section, bed D in, ...2213
" (Third) basin, Lower Freeport coal D in, ...2236
Westwood tunnel, Southern coal field, ...2113
Wetham coal basin, description of, ...2190-2
Weymer mine, bed C' in, ... 2254
Wharton coal bed in Eastern Middle field, ...2027 to 2047
" " basin, McKean county, ...1471
" basin of Potter county, equivalent to Caledonia basin,.....2280
" furnace property, section, ...1847
Wheatland, Mercer county, wells, depth of Corniferous in,..26-7,1190-1
White, David 2009; Deer mountain, 682-4-94, 871-3; Deer Valley, 1030, 1194, 1236; Deer valley quarries, 1004; Deer Hole valley, 875, 1262, 1332; mountain, 45, 871; mountain knob axis, 871; Oak mountains, Penn., 633; Oak colliery, 1963; Oak quarry, 594; R. pit, 2528; Rock mountain, ...248-9-51
White's coal bed or Lykens Valley No. 4 bed, Southern coal field, ..2136

INDEX FINAL SUMMARY REPORT. 95

White, Prof. I. C. general section, 1389, 1584; general section, Erie
    county, 1497; geology of Susquehanna river, 1409; col-
    ored county map, 1434; sections, 931, 1233, 1418-19, 1458;
    sections, Chemung in Erie county, 1492; Tioga, Sus-
    quehanna, Wayne, Bradford and Wyoming counties,
    1645; sections on the Lehigh, 1596-8; Stony Brook
    beds, 1564; measurement of No. IV, 631; classification
    of No. IX, 1576-7; collections on North Branch, Sus-
    quehanna, ...1009
Whitehall (Lehigh county) quarries, roofing slate, ...608-9
Whitely creek opening, Waynesburg coal in, ...2561
Whitfield's tunnel, Southern coal field; Buck mountain beds. 2097-9
Whitney, J. D. the Azoic system, ... 59
  " mine, Pittsburgh coal bed in, ...2486
Whittaker bank, Pittsburgh coal in, ...2540
Wiconisco anthracite basin, ...691,963,1811
Wigton mine, bed D in. 2257
Wilcox wells, Chemung rocks in 1482-5; Pocono shales in, ...1729-55
  " geyser, ...1486
  " Nos. 1 and 3, wells; records of, ...1486-7
Wild Cat mountain, 1255, 1567; range, 1275; ridge, 980, 1085,1213-21-69-72;
    pits, 245; valley, ...1560
Wilhelm mine section and analysis, Pittsburgh bed, ...2462
Wilkes-Barre division, Northern coal fields, ...1982
Wilkins opening, bed D in, ...2375
Will mine, bed D in, ...2256
Willcox, Col. Joseph notes on Delaware county minerals,............. 103
William A colliery, Northern coal field, ...1977
William Elliot tract bore hole, Natalie colliery, Western Middle
    coal field, ...2063-4
  " Penn basin, Mammoth vein, 2061; Catskill, ...1573
Williams mine, 1908, 2528; quarry, 1517-21; old slate quarry, 550-74;
    J. W., quarry, 588; David, quarry, 602; R. R. quarry, 604;
    Owen & Company's quarry, slate of No. III, ...602 to 606
Williamsburg gap, No. IV in, ... 661
  " section, Medina and Oneida rocks in, ... 667
  " valley, No. IV in, 684; Oriskany and Gaudagalli. 1143
Williamsport gap, Medina and Oneida rocks in, ... 667
Williamstown colliery, Southern coal field, ...2134 to 2138
  " system division, Southern coal field. 2133
Willow Grove, ...79 to 84,94,124,166,176,201
  " " Barren Hill, outcrop of Primal rocks. 175
  " " mine, Pittsburgh coal bed in, ...2542

96   GEOLOGICAL SURVEY OF PENNSYLVANIA.

Will's creek (Hyndman), section, Catskill rocks in, ... 1615
"   mountain, ... 847-8-9,997-8,1039,1281
"   "   gap, No. IV formation in, ... 649-63
"   "   anticlinal, Oriskany sandstone on, ... 1115
Wilmarth coal tract, Glen Mayo colliery, Alton lower coal, ... 1884
Wilmore (or First) Bituminous coal basin, ... 2158,2225-34-45-53-56
Wilmot anticlinal Chemung formation on, ... 1437
Wilson mine, bed E, ... 2340
"   fossil ore bed, Clinton rocks, ... 897
"   and Dewart tract, Western Middle coal field, ... 2069
Wind Gap, ... 584,1204-5,1568-9
Windy Harbor colliery, Southern coal field, ... 2109-11
Winfield quarries, No. VI limestone beds, ... 985
Wingert mine, bed D in, ... 2302
Winslow, Arthur, section of No. XII at Solomon's gap, ... 1984
"   section, No. XIII, Mt. Pisgah to Lehighton, ... 1595
Wire ridge, Genesee block slate in, ... 1327
Wishart mine, Barnet bed in, ... 2174
Wissahickon creek; geology of, ... 120 to 125
Wolf creek, mines on, Southern coal field, ... 2109 to 2114
Wolford's lime quarry, No. VI limestone in, ... 1116
Wolfsburg Kemble mines, fossiliferous shales of No. V at, ... 896
Woodbridge bank, section bed D, ... 2213
"   Theodore, warrant, coal beds in bore hole on, ... 1961
Woodbury-Curry anticlinal (Canoe Valley-Morrison's Cove), ... 409
Woodcock valley, ... 819-22,1574
Woodison mine, bed D in, ... 2332
Woodland fire clay analysis, ... 2216
Woodrow mines, bed E in, ... 2404
Woodside coal basin, Eastern Middle coal field, ... 2028
Woodvale shaft, Barnet bed in, ... 2183
Woodward colliery, Northern coal field, ... 1988 to 1996
Woods bank, Pittsburgh coal bed in, ... 2521
"   pit, Pittsburgh coal bed in, ... 2523
"   run mines, Pittsburgh coal bed in, ... 2554
Wolley's quarry section, No. VI limestone, ... 952
Woolsey bank, Sewickley bed in, ... 2524
Wright, bank, Quakertown coal in, 1911; Pittsburgh coal in, ... 2528
"   mine, Clarion coal bed in, ... 2390
Wrightsville and York limestone belt in Southern York county, ... 182
Wurtemburg section, Pottsville conglomerate in Beaver, ... 1915
Wyoming county, 1570-1-94; county, section, 1582; county, blue-stone
           quarries, 1591; basin, 1643, 1823-33; basin, depth of, 1949; re-
           gion, contents of, 2151; mountain, 1570, 1651; mountain, ter-

race, 1607; mountain, gap section, 1829; valley, 1947; colliery, 1990-5-6; shaft anticlinal, ...1936
Wythe county, Virginia, zinc ores, compared with Pennsylvania types, ... 437

Y.

Yager mine, limonite ore in, ... 233
Yard coal bed, Southern coal field, limonite mines, ... 2115
Yellow Breeches creek, limonite mines along, ... 246
" creek (Hopewell) section of Nos. IX and X rocks, ... 1615
" run shaft, bed B at, ... 2223
" Springs creek, deep shaft along, Peach mountain coal in, Southern coal field, ... 2104
" Springs gap, mine workings at, in Southern coal field, ... 2145
Yerger's falls, 1206; Yeager's road metal quarry, ... 1206
York, limonite ore banks north of, ... 214
" limonite ore banks west of, ... 215
" iron ore belt, ... 217
" county belt, 209; phyllite belt iron banks, 220; fossils, 2610, 2611; limestones and marbles of No. II, 473, 474; limestone valley of, 487; magnetic limonite mines, 256; ore banks, 208; limonite banks, 211; Peach Bottom slates, 114; Peach Bottom quarries, 193; quarries in the South mountains of, 556; South mountains of, 62, 70; white limestones and marbles of No. II in Chester, Montgomery and Centre counties. Chapter XXXIX, ... 467
" Farm colliery, Southern coal field, ...2105,2107,2116,2117
" Iron Company's mine, "Codorus ore" in, ... 223
" valley belt, Chickies quartzite in, 185; Sand valley new slate, 193; limestone, quarries south of the, 219; trap dykes, limestone belt, ... 454
Youghiogheny mines, Pittsburgh coal bed at, ... 2532
" No. 4 mine, Pittsburgh coal bed at, ... 2514
" river gap. Chemung flags in, 1500; Panama conglomerate in, ... 1511
" slope mine, Pittsburgh coal bed at, ... 2515
Young bank, bed B at, ... 2377
Young's drift, Southern coal field, ... 2142

Z.

Zerbe run gap, Western Middle coal field, Lyken's Valley coal, ... 2069
Zero coal bed, Southern coal field, ... 2135
Zimmerman bank, limonite ore in, ... 373
" mine, bed C' at, ... 2255

"   quarry, limestone of No. II in, ... 319
Zinc, in the New Red formation, ...2631
"   in formation No. II, Chapter XXXVII, ... 436
"   ores. Wythe county, Virginia, ... 437
"   ores of No. VI, Selinsgrove, ... 962
Zinn, Brinkley and, quarry, limestone of No. II. 312
Zollarsville coal bed, ...2578
Zuver well, Hosmer conglomerate in, ...1509

LIST OF PUBLICATIONS

OF THE

Second Geological Survey, 1874-1891.

Volume A, History of early geological explorations.
" AA, Anthracite region, report of progress. Panther creek basin.
" AA Atlases, 19 volumes of mine maps, topographical maps and sections of anthracite collieries; cross and columnar sections, etc.
    Northern coal field, six volumes.
    Eastern Middle coal field, three volumes.
    Western Middle coal field, three volumes.
    Southern coal field, six volumes (volume four in two parts).
    Anthracite statistics, 1883-4.
    New general map, Anthracite coal fields, 1890.
    Atlas, Carbon, Schuylkill, Dauphin and Berks Counties.
" A2. Report on Coal Waste in Mining Anthracite.
" AC. Mining Methods in the Anthracite region.
  AC. Atlas, Plates and illustrations to above.
" √B. Mineralogy of Pennsylvania, etc., by Dr. F. A. Genth.
" B2. Mineralogy of Pennsylvania, etc., by Dr. F. A. Genth (pamphlet sometimes bound in Volume B).
√ " C. York and Adams Counties, etc., with maps.
√ " C2. York, Adams, Cumberland and Franklin Counties, etc., maps.
" C3. Lancaster County.
" √ C3. Atlas Lancaster County, Maps.
" C4. Chester County, Map in pocket.
" C5. Delaware County, Map in pocket.
" C6. Philadelphia Belt. Philadelphia County, Southern Bucks and Southern Montgomery, Maps in pocket.

GEOLOGICAL SURVEY OF PENNSYLVANIA.

Volume C7. Atlas, parts of Bucks and Montgomery, Maps of water basins of Neshaminy, Tohickon and Perkiomen creeks, from surveys of the Philadelphia Water Department.
" D. Lehigh County, Brown Hematite ranges, with maps.
" D2. Lehigh Iron Ore district, Map of Lehigh County in pocket.
" D3, part 1. Report on Lehigh and Northampton County, the Roofing Slate Belt and Quarries. The Water Gaps. The Limestone Belt and Ore Mines, and the mountain topographical surveys.
" D3, part 2. Berks County.
" D3. Atlas, Maps of Northampton, Lehigh and Berks Counties. Durham and Reading Hills. Iron Ore Mines.
" D5. Atlas, South Mountain sheets. Maps of Adams, Franklin and Cumberland Counties.
" D6. Atlas, South Mountain Sheets (continued).
" E. Azoic Rocks of South Eastern Pennsylvania.
" F. Fossil Ores of the Juniata Valley, Aughwick Valley and East Broad Top, Maps in pocket.
" F2. Perry County, Map in pocket.
" F3. Union, Snyder, Mifflin and Juniata Counties, Maps in pocket.
" F3. Atlas Union, Snyder, Huntingdon, Centre and Juniata Sectional Maps, Seven Mountains.
" G. Bradford and Tioga Counties. Maps in pocket.
" G2. Lycoming and Sullivan Counties. Maps in pocket.
" G3. Potter County. Map in pocket.
" G4. Clinton County. Map in pocket.
" G5. Susquehanna and Wayne Counties. Maps in pocket.
" G6. Pike and Monroe Counties. Maps in pocket.
" G7. Wyoming, Lackawanna, Luzerne, Columbia, Montour and Northumberland Counties outside of the Anthracite Coal Fields. Maps in pocket.
" H. Clearfield and Jefferson Counties. Maps in pocket.
" H2. Cambria County, with maps.
" H3. Somerset County, with maps.
" Atlas to H2 and H3. Maps of Cambria and Somerset Counties; pamphlet.
" H4. Indiana County. Map in pocket.
" H5. Armstrong County. Map in pocket.
" H6. Jefferson County. Map in pocket.
" H7. Clearfield County (revised) 1883. Map in pocket.
" I. Oil Region. Venango District, with map.
" I.2. Oil Region. Well records.
" I.3. Oil Region. Warren, Venango, Clarion and Butler Counties. Oil well rig and tools. Ancient drainage.
" Atlas to I.3. Oil Region., Maps and Charts.
" I.4. Oil Region. Warren County. Maps in pocket.

Volume I.5 Oil and Gas Fields, Western Pennsylvania. Maps in pocket.
" J. Special Report on Petroleum.
" K. Greene, Washington and Southern Allegheny Counties, Bituminous Coal Felds. Maps in pocket.
" K2. Eastern Allegheny County, Fayette and Westmoreland Counties, West of Chestnut Ridge. Maps in pocket.
" K3. Ligonier Valley. Maps in pocket.
" K4. Monongahela River Mines and Lower Youghiogheny. Maps.
" L. Coke Manufacture. Methods of Coking. Use of Natural Gas in Iron Manufacture, with illustrations.
" M. Chemical analyses of Coals, Ores, Limestone, etc.
" M2. Chemical Analyses.
" M3. Chemical Analyses.
" N. Levels Above Tide.
" O. Museum Catalogue.
" O2. Museum Catalogue.
" O3. Museum Catalogue.
" P. Coal Flora Text, Volumes 1 and 2.
" P. Coal Flora Text, Volume 3 and Plates.
" P. Coal Flora, Atlas of Plates.
" P2. Permian Flora.
" P3. Ceratiocaridae and Eurypteridae, with plates.
" P4. Dictionary of Fossils, 3 volumes.
" Q. Beaver, North Allegheny and South Butler Counties. Maps in pocket.
" Q2. Lawrence County and the Ohio Line. Map in pocket.
" Q3. Mercer County. Map in pocket.
" Q4. Erie and Crawford Counties. Maps in pocket.
" R. McKean County.
" R. Atlas, McKean County maps and charts.
" R2. Cameron, Elk and Forest Counties.
" R2. Atlas, maps and charts to above.
" T. Blair County.
" T. Atlas, Blair County, topographical sheets of Morrison's Cove.
" T2. Bedford and Fulton Counties. Maps in pocket.
" T3. Huntingdon County. Map in pocket.
" T4. Centre County. Maps in pocket.
" V. Northern Butler, Special Survey along Beaver and Shenango Rivers. Maps in pocket.
" V2. Clarion County. Maps in pocket.
" X. Geological Atlas of Counties. Scale of maps, 6 miles to 1 inch.
" Z. Terminal Moraine.
Annual Report, 1885, one volume text and one volume atlas, contains:
    1. Oil and Gas.
    2. Vegetable Origin of Coal.

3. Pittsburgh Coal Region.
4. Wellsburg Coal basin.
5. Tipton Run Coal Basin.
6. Anthracite Coal Region.
7. Wyoming County Fossils.
8. Bernice Coal Basin.
9. Mehoopany Coal Field.
10. Cornwall Ore mines.
11. Delaware and Chester Kaolins.
12. Quarternary Geology, Wyoming Valley.
13. Pressure, etc., of Rock Gas.
14. Progress Geodetic Survey.

Annual Report, 1886, four volumes text and two volumes atlas, contains:
    i. Pittsburgh Coal Region.
    ii. Oil and Gas Regions.
    iii. Anthracite Coal Region with atlas.
    iv. 1. The Lehigh River Cross Section.
        2. Paint Ores along the Lehigh River.
        3. Iron Ore Mines and Limestone Quarries of the Cumberland-Lebanon Valley.
        4. Geology of Radnor Township, Delaware County, with an atlas.

Annual Report, 1887, one volume with maps, contains:
    1. Cave Fossils.
    2. Fossil Tracks in the Trias.
    3. New Boston Anthracite Basin.
    4. State Line Serpentine.

Grand Atlases.

These atlases contain maps and sections printed on heavy paper (sheets 26 inches by 32 inches) of similar maps and sections printed on light paper sheets contained in the octavo atlases.

Division I, Part 1, Geological maps of 56 counties.
Division II, Part 1, Anthracite Coal Fields.
Division II, Part 2, Anthracite Coal Fields.
Division III, Part 1, Petroleum and Bituminous Coal Fields.
Division IV, Part 1, Durham and Reading Hills and South Mountains.
Division V, Part 1, Central and Southeastern Pennsylvania.

Appendix to Grand Atlas, edition in rolls, containing three sheets of South Mountain topography, a geological map of Berks county and of Huntingdon county.

Summary Final Report, Geology of Pennsylvania. Vol. I, describing Laurentian, Huronian, Cambrian and Lower Silurian formations and their sub-divisions.

Summary Final Report, Geology of Pennsylvania. Vol. II, describing Upper Silurian and Devonian formations and sub-divisions.

Summary, Final Report, Geology of Pennsylvania, Vol. III, two parts:
Part I. Carboniferous rocks and Anthracite report.
Part II. Bituminous Coal Fields and New Red of Bucks and Montgomery Counties.

Atlas to Final Summary Report, contains a new geological map of the State, scale 6 miles to 1 inch; map of the Bituminous coal district, scale 4 miles to 1 inch.

Geological map of Lebanon county, geological and topographical maps and a section of Bucks and Montgomery Counties.

A BRIEF GUIDE TO THE PUBLICATIONS OF THE PENNSYLVANIA GEOLOGICAL SURVEY.

1874-1891.

A.

Adams county, with map, Vol. C.
Adams and Franklin counties, Vol. C2.
Adams county, South mountain maps, Atlases D5 and D6.
Adams county, geological map, Atlas D5.
Adamsville-Jamestown section, Atlas to Vol. I.3.
Alden shaft, cross section, Vol. AA, Northern coal field, Part 5.
Allegheny county, see geological map Southwestern Pennsylvania, Annual Report, 1886, Part 1.
Allegheny county, eastern part, map, Vol. K2.
Allegheny county, northern part, map, Vol. Q.
Allegheny county, southern part, map, Vol. K.
Allegheny county, oil wells, Vol. I.4.
Allegheny mountain, Bennington cross section, Atlas H2 and H3.
Altamount-Pottsville gap cross section, Atlas AA, Southern coal field, Part 5.
Alton coal basin, map, Vol. R, Atlas.
Analyses of minerals, Vols. B, B2.
Analyses of coals, ores, limestones, etc., Vols. M, M2, M3.
Analyses of limestones and ores, Vol T4.
Analyses, ancient water course near Franklin, Atlas to Vol. I.3.
Anthracite region, general map, AA Atlas, Southern coal field, Part 1. General map, Annual Report 1886, Part 3, Atlas.
Anthracite region, report, Part 1, Vol. AA, Part 2, Annual Report, 1885, Part 3, Annual Report 1886, Part 2; report, Annual Report 1886, Part 3 and Atlas.
Anthracite region report, summary final report, Vol. III. Part 1.
Anthracite tonnage, 1870-1882, Vol. AA.
Anthracite tonnage, statistical sheet, AA Atlas, Southern coal field, Part 1.
Anthracite tonnage for 1885 and 1886, Annual Report 1886. Part 3, Atlas.
Anthracite coal beds, estimate of contents, Vol. AA.
Anthracite columnar section, Vol. AA.

7*

Anthracite maps, Atlas AA, Northern coal field, Parts 1 to 6; Eastern Middle, Parts 1 to 3; Western Middle, Parts 1 to 3; Southern, Parts 1 to 6.
Anthracite, methods of mining and appliances, Vol. AC and Atlas.
Anthracite operators, list of, Vol. AA.
Anthracite rocks below the coal, Vol. G7.
Anthracite slack, coking with bituminous coal, Vol. M2.
Anthracite, waste in mining, Vol. A2.
Antrim coal field, Vol. G.
Anchor colliery, section, Atlas AA, Southern Anthracite field, Part 4.
Anchor-Fogarty cross section, AA, Southern Anthracite field, Part 5.
Andrew B. White tunnel section, AA, Southern Anthracite field, Part 4.
Archaen rocks of the State, described, Fin. Sum. Rep., Vol. I.
Armstrong county, with map, Vol. H5.
Armstrong, Butler and Clarion oil field, map, Atlas to Vol. I.3.
Armstrong, Butler and Clarion levels, map and profile, Vol. J.
Archbald pot holes, Ann. Rep. 1885.
Archbald mine sheets and sections, Atlas AA, Northern field, Part 4.
Arnot coal field, Report G.
Ashland and vicinity, maps and section, Atlas AA, Western Middle, Part 1.
Ashley mine sheets and sections, Atlas AA, Northern field, Part 1.
Aughwick valley, Vol. F.
Azoic rocks, Vol. E.

B.

Bald Eagle mountain faults, Vol. T3.
Bald Eagle mountain and Nittany valley maps, Vol. T3.
Ball's colliery cross section, Atlas AA, Southern field, Part 5.
Barclay coal field, Vol. G.
Bearings, distances and levels in York and Adams counties, Vol. C.
Bear Ridge—Windy Harbor cross section, Atlas AA, Southern field, Part 5.
Bear Valley basin, map and sections, Atlas AA, Western Middle, Part 2.
Beaver county, North Allegheny and South Butler, with maps, Vol. Q.
Beaver county oil wells, Vol. I.4.
Beaver and Shenango rivers, Slippery Rock creek, map, Vol. V.
Beaver Valley, Vol. V.
Beaver Meadow basin, Carbon county, maps and sections, Atlas AA, Eastern Middle, Part 2.
Bedford and Fulton counties, with maps, Vol. T2.
Bedford red beds of the Crawford shales, underground map, Vol. I.5.
Bedford Springs, analysis of water from, Vol. T2.
Beechwood colliery, Col. section, AA Atlas, Southern field, Part 4.
Beechwood colliery, cross section, AA Atlas, Southern field, Part 5.

INDEX FINAL SUMMARY REPORT.

Belgian and Bee-Hive coke ovens compared, Vol. M2, p. 259.
Bell's colliery section, AA Atlas, Southern field, Part 4.
Belmont colliery, Col. section, AA Atlas, Northern field, Part 5.
Belmont colliery, cross section, AA Atlas, Northern field, Part 6.
Beltheiser colliery, section, AA Atlas, Southern field, Part 4.
Bennington cross section, Atlas H2 and H3.
Berks county report, geology of, Vol. D3, Part 2.
Bernice coal basin, Ann. Rep. 1885.
Bibliography of petroleum, Ann. Rep. 1886, Part 2.
Big Diamond colliery, cross section, AA Atlas, Southern field, Part 5.
Big Flats mine sheet, AA Atlas, Southern field, Part 3.
Bitumnious coal field, geology of, Sum. Final Report, Vol. III, Part 2.
Black Creek basin, Luzerne county mine maps and sections, AA Atlas, Eastern Middle, Part 1.
Black Creek and Priscilla Lee section, AA Atlas, Northern field, Part 5.
Black mine, Richardson cross section, AA Atlas, Southern field, Part 5.
Blackwood colliery, mine sheet, AA Atlas, Southern field, Part 2.
Blackwood colliery, Col. section, AA Atlas, Southern field, Part 4b.
Blackwood—Otto cross section, AA Atlas, Southern field, Part 6.
Black Rock, N. Y., to Dunkard creek, cross section, Vol. I3, Atlas.
Blair county, coal measures, Vol. H2.
Blair county, palaeozoic section, Vol. F.
Blair county, iron industries, Vol. T.
Blair county, general geology, Vol. T and Atlas.
Blair county, Tipton run coals, Ann. Rep. 1885.
Blossburg coal field, Vol. G.
Bore hole records, Marvine, Grassy Island, Marshwood, Jermyn, Eddy's Creek, Olyphant,, Eddy Island, Mt. Jessup, Pierce Creek, Eaton, Glenwood, etc., etc., AA Atlas, Northern field, Part 5.
Bore hole sections near Frackville and Gordon, AA Atlas, Southern field, Part 4.
Border rocks, anthracite measures, Vol. G7.
Bowman's colliery section, AA Atlas, Southern field, Part 4.
Boyd's Hill well record compared with Leechburg well, Vol. L.
Boyertown ore mines, map, Vol. D3 Atlas.
Brace Brook mine sheets and sections, AA Atlas, Northern field, Part 4.
Bradford county, Vol. G.
Bradford oil districts, Vol. R and Atlas.
Bradford oil district, sub-conglomerate rocks, Vol. R. and atlas.
Bradford oil district, statistics of production, Atlas to Vol. R.
Bradford oil sand, relation to lower carboniferous, Atlas to Vol. R.
Branchdale mine sheet, AA Atlas, Southern field, Part 2.
Broad Top coal field in Bedford and Fulton counties, Vol. T2.
Broad Top coal field in Huntingdon county, Vol. T3.
Broad Top coal field, general review of Final Summary Report, Vol. III, Part 2.

GEOLOGICAL SURVEY OF PENNSYLVANIA.

Broad mountain basin, West End mine sheet, AA, Atlas, Southern field, Part 2.
Broken Straw valley, Warren county, topographical map, Atlas to Vol. I3.
Brookside colliery mine sheet, AA Atlas, Southern field, Part 3.
Brookside colliery, cross sections, AA Atlas, Southern field, Part 6.
Brown hematite ore ranges, Lehigh county, D, D2, M3, preface.
Bucks county, southern part, map, Vol. C6.
Bucks county, central part, map, Vol. C7, Atlas.
Bucks county, New Red measures in, and Montgomery, Fin. Sum. Rep. Vol. III, Part 2.
Buckville colliery section, AA Atlas, Southern field, Part 4.
Buffalo Coal Company, map, Atlas Vol. R.
Building stone quarries in Ohio, Vol. I.3.
Buried river valleys, Vol. Q2, p. xii, Ann. Rep. 1885.
Butler county, southern part, North Allegheny and Beaver, maps, Vol. Q.
Butler county, northern part, Vol. V.
Butler county, oil region and wells, Vols. I.3 and I.4.
Butler county oil well records, Vol. V.
Butler, Armstrong and Clarion oil fields, map, Atlas to Vol. I.3.
Butler, Armstrong and Clarion line of levels, map profile, Vol. J.

C.

Cameron, Elk and Forest counties, Vol. RR and Atlas.
Cameron Columnar sections, RR Atlas.
Cameron oil wells, Vol. I.4.
Cambria county and maps, Atlas H2 and H3.
Cambria and Somerset counties, Vols. H2 and H3.
Cameron oil wells, Vol. I.4.
Cameron columnar sections, Atlas H2 and H3.
Cambrian rocks of Europe and America, Vol. E.
Cambrian rocks in the State, described, Fin. Sum. Rep., Vol. I.
Carbon determination in iron and steel. Vol. M2. p. 409.
Carbondale mine sheets and sections, AA Atlas, Northern anthracite, Part 4.
Carboniferous rocks (Pocono, Mauch Chunk and conglomerate), described, Fin. Sum. Rep., Vol. III, Part 1.
Carboniferous in West Virginia, Vol. PP.
Carboniferous along West Branch of Susquehanna, Vol. G4.
Catalogue of specimen in museum, Vols. O, O2, O3.
Catalogue of specimens in Berks county. Vol. D3, Part 2.
Catalogue of specimens in Adams, Franklin and Lancaster, Vol. C3.
Catalogue of specimens in Huntingdon county, Vol. F.
Catalogue of specimens in York and Adams counties. Vols. C, C2.
Cave fossils, Ann. Rep. 1887.
Cayuga colliery, cross section, AA Atlas, Northern anthracite, Part 6.

Cements, analyses and tensile strength, Vol. D. 2.
Cement works in Lehigh county, Vol. D2.
Centre county, with map, general geology of, Vol. T4.
Central colliery cross section, AA Atlas, Northern anthracite, Part 6.
Ceratiocaridae, with plates, Vol. P.3.
Chamberlain colliery section, AA Atlas, Northern Anthracite, Part 4.
Chamberlain colliery, cross section, AA Atlas, Northern anthracite, Part 6.
Charcoal iron furnaces, Clarion county, Vol. V2.
Cherry Tree gas well, Vol. H2.
Chester county, with map, Vol. C4.
Chester county, State line Serpentine, Ann. Rep. 1887.
Chester and Delaware Kaolin deposits, Ann. Rep. 1885.
Chestnut Hill iron ore banks, map, C3 Atlas.
Clarion county, with map, Vol. V2.
Clarion county, warrantee map, RR Atlas.
Clarion county, oil wells, 1.3, 1,4.
Clarion, Armstrong and Butler levels, map profile, Vol. J.
Clarion, Armstrong and Butler oil fields, 1.3, Atlas.
Clarion coal bed split, Vol. V.
Clarke colliery, cross section, AA Atlas, Northern anthracite, Part 6.
Clarge and McCormick cross section, AA Atlas, Southern anthracite, Part 6.
Clearfield and Jefferson counties, Vol. H.
Clearfield county, revised, with map, Vol. H7.
Clinton county, Vol. G4.
Clinton county oil wells, Vol. I.4.
Coal formation, by H. D. Rogers, Vol. H, appendix.
Coal, clasification of, Vol. M2.
Coal, genesis of, Vol. H4, preface.
Coal vegetable origin of, Ann. Rep. 1885.
Coal, structure, troubles, clay veins, Vol. K3.
Coal beds, identification of, Vol. Q2, p. xxi.
Coal, bituminous and anthracite, occurring together, Vol. G2.
Coal, anthracite, geology of, Fin. Sum. Report, Vol. III, Part 1.
Coal, bituminous, geology of, Fin. Sum. Report, Vol. III, Part 2.
Coal measures in Western Pennsylvania and Eastern Ohio, correlation of, Vol. Q2.
Coal flora, text, P, Vols. 1, 2 and 3, Atlas P.
Coal waste in mining anthracite, Vol. A2.
Coal Brook colliery, columnar section, AA Atlas, Northern Anthracite, Part 5.
Coal Brook colliery, cross section, AA Atlas, Northern anthracite, Part 6.
Coal Hill, Milford cross section, AA Atlas Southern anthracite, Part 5.
Coke manufacture, special reports, G. L.
Coking coal best, Vol. G.

Coking qualities of coal, Vol. G2.
Cold Springs mine sheets, AA Atlas, Southern anthracite, Part 3.
Colket colliery, columnar section, AA Atlas, Southern anthracite, Part 4.
Colket colliery, cross section, AA Atlas, Southern anthracite, Part 6.
Columbia county, Vol. G7.
Columnar sections, anthracite, Vol. AA.
Columnar sections from No. VIII to No. XVII, Vol. I.5.
Columnar sections from top of Upper Barren measures, Vol. I.5, Atlas.
Columnar sections, Oil City, Franklin, Cranberry coal bank, Vol. I.3.
Comparative geology of Northeastern Ohio, Northwest Pennsylvania and Western New York, Vol. I.
Connellsville coke trade, Vol. L.
Conglomerate, discussion of its formation, Vol. H5, preface.
Conglomerate, Pottsville formation, geology of, Fin. Sum. Rep., Vol. III, Part 1.
Conglomerate, Garland and Panama, Vol. I.3.
Contents anthracite coal beds, estimated, Vol. AA.
Continental colliery cross section, AA Atlas, Northern anthracite, Part 6.
Contorted beds of micaceous gneiss, Philadelphia belt, Vol. C5, p. x.
Copper ores in the United States, Vol. C3.
Cornwall ore mines, report and map, Ann. Rep. 1885.
Cornwall ore mines, history and present ownership, Ann. Rep. 1885.
Corniferous limestone, underground map, Vol. I.5.
Correlation of coal measures in Western Pennsylvania and Eastern Ohio, Vol. Q2.
Cost of survey from 1874 to 1885, Ann. Rep. 1885, p. xxi.
Counties, summary geology of, Vol. X.
Counties, Atlas, colored geological maps of all counties, Vol. X.
Crawford and Erie counties, with maps, Vol. Q4.
Crawford county oil wells, Vol. I.4.
Cumberland county, South mountain maps. D5 and D6, Atlases.
Cumberland county, geological map, Atlas, D5.
Cumberland and Lebanon Valleys, iron ores and limestones, Ann. Rep. 1886, Part 4 and Atlas.
Cumberland Valley iron ores, map and analyses, Vol. M3.

D.

Dauphin mine sheet, AA Atlas, Southern anthracite, Part 3.
Delano to Mahanoy City, basin, maps and sections, AA Atlas, Western Middle, Part 1.
Delaware county, with map, Vol. C5.
Delaware and Chester county Kaolin deposits, Ann. Rep. 1885.
Delaware county, Radnor township, notes on geology, Ann. Rep. 1886, Part 4 and Atlas.
Delaware water gap, Vol. G6.

Deringer, Black Creek basin, maps and sections, AA Atlas, Eastern Middle, Part 3.
Devonian along West Branch of Susquehanna county, G4.
Devonian and Upper Silurian rocks of the State, described, Fin. Sum. Rep., Vol. II.
Diamond colliery section, AA Atlas, Southern anthracite, Part 4.
Diamond colliery, cross section, AA Atlas, Southern anthracite, Part 5.
Dictionary of fossils, P4, Vols. I, II and III.
Dillsburg ore bed, preliminary discussion on, Vol. C2.
Dip of strata, method of determining, Vol. V, p. vi.
Distances, levels, etc., in York and Adams counties, Vol. C.
Derkes colliery, section, AA Atlas, Southern anthracite, Part 4.
Dolomites, limestones, notes on, M3.
Dodge colliery, cross section, AA Atlas, Northern anthracite, Part 6.
Donaldson mine sheet, AA Atlas, Southern anthracite, Part 3.
Drainage of Erie county, pre and post glacial, Vol. I.3.
Drifton, Highland and mine maps and sections, AA Atlas, Eastern Middle, Part 1.
Dundas No. 6 colliery, section, AA Atlas, Southern anthracite, Part 4.
Dunkard creek Pa., to Black oRck, N. Y., cross section, Vol. I.3, Atlas.
Dunmore colliery, mine sheets and sections, AA Atlas, Northern anthracite, Part 3.
Dunn colliery, cross section, AA Atlas, Northern anthracite, Part 6.
Dupont and Henderson cross section, AA Atlas, Northern anthracite, Part 5.
Durham and Reading Hills, maps, eighteen sheets, Vol. D3, Atlas.

E.

Eagle Hill colliery section, AA Atlas, Southern anthracite, Part 4.
Eagle Hill colliery, cross section, AA Atlas, Southern anthracite, Part 5.
East Broad Top district, Vol. F and Fin. Sum. Rep., Vol. 3, Part 2.
Eastern Middle anthracite mine maps, basins of Upper Lehigh, Pond Creek, Sandy Run, Little Black creek, Big Black creek, Hazle creek, Cranberry creek, Beaver creek and Green mountain, AA Atlas, Eastern Middle, Parts 1, 2 and 3.
East Franklin colliery, columnar section, AA Atlas, Southern anthracite, Part 4b.
Eaton colliery, columnar section, AA Atlas, Northern anthracite, Part 5.
East Pine Knot, York Farm cross section, AA Atlas, Southern Anthracite, Part 5.
Edgerton colliery, cross section, AA Atlas, Northern anthracite, Part 6.
Elevations in Warren county, Vol. I.4.
Elk, Cameron and Forest counties, maps. Vol. R2 and Atlas.
Elk county, warrantee map, R2, Atlas.

Elk county oil wells, Vol. I.4.
Ellendale Forge colliery mine sheet, AA Atlas, Southern anthracite, Part 3.
Ellendale, North vein drift section, AA Atlas, Southern anthracite, Part 6.
Ellsworth colliery section, AA Atlas, Southern anthracite, Part 4.
Erie and Crawford counties, with maps, Vol. Q4.
Erie county oil wells, Vol. I.4.
Erie colliery, columnar section, AA Atlas, Northern anthracite, Part 5.
Erie colliery cross section, AA Atlas, Northern anthracite, Part 6.
Erie, lake, glacial drainage, Vol. I.3.
Erosion, chemical, Nittany valley limestones, Vol. T3.
Erosion, by solution, Vol. T4, p. 420.
Erosion, rate of, in Allegheny Valley, Vol. V2, p. ix.
Erosion, map of Titusville and Pleasantville, Vol. I.3, Atlas.
Eurypteridae, with plates, Vol. P3.
Exeter shaft, colliery, cross section, AA Atlas, Northern anthracite, Part 5.

F.

Fall Brook coal field, Vol. G.
Fayette county, with map, Vols. K2 and K3.
Fayette and Westmoreland, Ligonier valley, Vol. K3.
Feger ridge colliery, cross section, AA Atlas, Southern anthracite, Part 6.
Ferriferous limestone, general, Vol. H3.
Ferriferous limestone, limit of, Vol. H4.
Ferriferous limestone, Armstrong county, Vol. H5, preface.
Fire brick tests, Vol. M2.
Fire clays, genesis of, Vol. Q2, p. xix.
Fire clays of Clearfield county, Vol. H7, appendix.
Fishing creek, mine sheet, AA Atlas, Southern anthracite, Part 3.
Fogarty-Anchor cross section, AA Atlas, Southern anthracite, Part 5.
Forty Fort cross section, AA Atlas, Northern anthracite, Part 5.
Fort Lookout mine sheet, AA Atlas, Southern anthracite, Part 3.
Fort Lookout shaft section, AA Atlas, Southern anthracite, Part 4.
Forest, Elk and Cameron counties, maps, Vol. R2 and Atlas.
Forest county, warrantee map, R2, Atlas.
Forest county oil wells, Vol. I.4.
Forest City mine sheets and sections, AA Atlas, Northern anthracite, Part 4.
Forest City bore holes and columnar sections, AA Atlas, Northern anthracite, Part 5.
Forestville colliery mine sheet, AA Atlas, Southern anthracite, Part 2.
Forestville colliery columnar section, AA Atlas, Southern anthracite, Part 4b.

INDEX FINAL SUMMARY REPORT.   XV

Forestville colliery, cross sections, AA Atlas, Southern anthracite, Part 5.
Forestville deep shaft, section, AA Atlas, Southern anthracite, Part 4.
Formation, Nos. 8 to 17, columnar sections, Vol. I.5.
Former surveys, history of, Vol. A.
Fossils of Southwestern Pennsylvania, list of, Vol. K3.
Fossil dictionary, P4, Vols. 1, 2 and 3.
Fossils found in caves, Ann. Rep. 1887.
Fossil fungus, description of, Vol. Q.
Fossil ores, Juniata valley, Vol. F.
Fossil tracks in Trias. Ann. Rep., 1887.
Frackville, mine sheet, AA Atlas, Southern anthracite, Part 2.
Frackville, bore hole sections, AA Atlas, Southern anthracite, Part 4.
Franklin county, Vol. C2.
Franklin county, South mountain maps, D5, D6, Atlas.
Franklin county, Geological map, Atlas D5.
Fritz Island ore mines, map, D3, Atlas.
Fulton county, with map, Vol. T2.

G.

Gaines coal field, Vol. G.
Gas and oil reports, Vols. I, I.2, I.3, I.4, I.5, Ann. Rep., 1885.
Gas and oil reports, Ann. Rep., 1886, Part 2.
Gas, natural, analyses, Vol. L.
Gas, natural, composition and fuel value, Ann. Rep., 1886, Part 2.
Gas, natural, pressure quantity and fuel value, Ann. Rep., 1885.
Gas, natural, use in iron working, Vol. L.
Gas, natural, durability of supply, Vol. L.
Gas coal trade of the Youghiogheny, Vol. L.
Gas well, Cherry Tree, Vol. H2.
Gap Nickel mine map, Vol. C3 and Atlas.
Garland conglomerate, Vol. I.3.
Genesis of iron ores, Vol. T4, p. 407.
Genesis of pipe ore limonite, Vol. Q2, p. xvii.
Geodetic triangulation, Ann. Rep., 1885.
Geological Hand Atlas of counties, Vol. X.
Geological structure of Pennsylvania, Vol. X.
Geology of Southeastern Pennsylvania, introduction, Vol. E.
Geology, comparative of Northeastern Ohio, Northwestern Pennsylvania and Western New York, Vol. I.
Girardville Collieries, maps and sections, AA Atlas. Western Middle, Part 1.
Glacial theories, Vol. Z.
Glacial drainage, Erie county, Vol. I.3.
Glacial formations, Pike county, Vol. G6.
Glacial deposits, Lehigh county, Vol. D2.
Glacial terminal moraine, Vol. Z.

Glaciated Region in Pennsylvania, Vol. Z, map.
Glen Carbon, mine sheet, AA Atlas, Southern anthracite, Part 2.
Glendower colliery, section, AA Atlas, Southern anthracite, Part 4.
Glenwood colliery, column, section, AA Atlas, Northern anthracite, Part 5.
Glenwood colliery, cross section, AA Atlas, Northern anthracite, Part 6.
Gold mine colliery, mine sheet, AA Atlas, Southern anthracite, Part 3.
Gold mine colliery, gap section, AA Atlas, Southern anthracite, Part 4.
Good Spring colliery, mine sheet, AA Atlas, Southern anthracite, Part 3.
Good Spring colliery, columnar section, AA Atlas, Southern anthracite, Part 4b.
Good Spring colliery cross section, AA Atlas, Southern anthracite, Part 6.
Gordan mine sheet, AA Atlas, Southern anthracite, Part 2.
Gordon bore hole sections, AA Atlas, Southern anthracite, Part 4.
Gorman's colliery section, AA Atlas, Southern anthracite, Part 4.
Gowen colliery, maps and sections, AA Atlas, Eastern Middle, Part 3.
Grassy Island colliery, columnar section, AA Atlas, Northern anthracite, Part 5.
Grassy Island colliery, cross section, AA Atlas, Northern anthracite,, Part 6.
Gratz tunnel, section, AA Atlas, Southern anthracite, Part 4.
Greene county, report and map, Vol. K and Ann. Report, 1886.
Greene county oil wells, Vol. I.4.
Greene and Washington bituminous district maps, Vol. K and Atlas, Ann. Rep., 1886.
Greene mountain basin, maps and sections, AA Atlas, Eastern Middle, Part 3.
Greenwood tunnel section, AA Atlas, Southern anthracite, Part 4.

H.

Hacupton colliery, cross section, AA Atlas, Northern anthracite, Part 6.
Hanover colliery, cross section, AA Atlas, Northern anthracite, Part 5.
Harris colliery, section, AA Atlas, Southern anthracite, Part 4.
Hazen's coal bank, Miller's quarry section, Vol. I.3, Atlas.
Hazle creek basin, Luzerne county, mine maps and sections, AA Atlas, Eastern Middle, Part 1.
Heckscherville mine sheet, AA Atlas, Southern anthracite, Part 2.
Hematite ores, genesis of, Vol. D2.
Hematite ores of Cumberland valley, M3.
Hematite ores, mining and washing, Vol. D.
Hematite ores, Lehigh county, Vol. D.
Herbine colliery section, AA Atlas, Southern anthracite, Part 4.
Herbine colliery cross section, AA Atlas, Southern anthracite, Part 5.
Hickory colliery, cross section, AA Atlas, Southern anthracite, Part 5.

Hillman colliery, cross section, AA Atlas, Northern anthracite, Part 5.
Hill's tunnel section, AA Atlas, Southern anthracite, Part 4.
History, previous surveys, Vol. A.
History of survey, 1874 to 1885, Ann. Rep., 1885, p. xvii.
Hollenback colliery cross section, AA Atlas, Northern anthracite, Part 5.
Homewood-Sharon section, map, Vol. V.
Honeybrook basin, Carbon county, map and sections, AA Atlas, Eastern Middle, Part 3.
Humboldt map and sections, AA Atlas, Eastern Middle, Part 3.
Huntingdon county, with map, Vol. T3.
Huntingdon county, cross section, Vol. F.
Huntingdon county, carboniferous, Vol. F.
Hyde Park mine sheets and sections, AA Atlas, Northern anthracite, Part 3.

I.

Ice sheet, its thickness, Vol. Z.
Identification of beds, anthracite, Vol. AA.
Indian relics, Berks county, Vol. D3, Part 2.
Indian sculptures, Vol. C3, Part 2.
Indiana county, with map, Vol. H4.
Iron furnaces in Bedford county, Vol. T2.
Iron industry, history of, Vol. D3, Part 2.
Iron industry of Centre county, Vol. T4.
Iron industry of Chester county, Vol. C4.
Iron ores, genesis of, Vol. T4, p. 407.
Iron ores, mining methods, Vol. T4.
Iron ores in caverns, Vol. T4, p. 418.
Iron ores of Cumberland valley, map and analyses, Vol. M3.
Iron ore mines Cumberland and Lebanon valleys, Ann. Rep., 1886, Part 4, Atlas.

J.

Jamestown, Adamsville section, Vol. I.3, Atlas.
Jefferson county, with map, Vol. H6.
Jefferson county oil wells, Vol. I.4.
Jefferson district, Vol. H.
Jermyn colliery, mine sheets and sections, AA Atlas, Northern anthracite, Part 4.
Jermyn colliery, cross sections, AA Atlas, Northern anthracite, Part 6.
Jessup colliery, mine sheet and sections, AA Atlas, Northern anthracite, Part 4.
Jones iron ore mine map, Vol. D2, Atlas.
Juniata county, with map, Vol. F3 and Atlas.
Juniata valley fossil ores, Vol. F.

K.

Kaolin, Delaware and Chester counties, Ann. Rep., 1885.
Kalmia colliery mine sheet, AA Atlas, Southern anthracite, Part 3.
Kalmia colliery, columnar section, AA Atlas, Southern anthracite, Part 4b.
Kalmia colliery cross section, AA Atlas, Southern anthracite, Part 6.
Kaska William colliery section, AA Atlas, Southern anthracite, Part 4.
Kaska William colliery cross section, AA Atlas, Southern anthracite, Part 5.
Kemble-Red mountain cross section, AA Atlas, Southern anthracite, Part 6.
Kentucky colliery section, AA Atlas, Southern anthracite, Part 4.
Kentucky colliery cross section, AA Atlas, Southern anthracite, Part 5.
Keystone colliery columnar section, AA Atlas, Northern anthracite, Part 5.
Kingston mine sheets and sections, AA Atlas, Northern anthracite, Part 1.
Kohler's gap mine sheet, AA Atlas, Southern anthracite, Part 3.

L.

Laboratory analyses, Vols. M, M2 and M3.
Lackawanna county, Vol. G7.
Lackawanna mine sheets and sections, AA Atlas, Northern anthracite, Part 2.
Lackawanna valley, Quarternary geology, Ann. Rep., 1885.
Lackawanna valley, topographical map of part, 2 sheets, Ann. Rep., 1886. Part 3, Atlas.
Lake Erie, pre-glacial outlet, Vol. Q4.
Lancaster county and map, Vol. C3 and Atlas.
Lancaster section, Neffsville, Marticville, C3, Atlas.
Latitude of Wilkes-Barre, Vol. AA.
Laurentian rocks described, Fin. Sum. Rept., Vol. I.
Lawrence county, with maps, Vol. Q2.
Lebanon county, Cornwall Mines, Ann. Rep., 1885.
Lebanon valley iron ores and limestone, Ann. Rep., 1886, Part 4 and Atlas.
Leggitt's creek colliery cross section, AA Atlas, Northern anthracite, Part 6.
Lehigh county, Vols. D, D3, Part 1.
Lehigh and Northampton counties, map, Vol. D3, Atlas.
Lehigh district and map. Vol. D2.
Lehigh Navigation Company's land report, Vol. AA.
Lehigh river cross section, Ann. Rep., 1886, Part 4 and Atlas.
Lehigh river paint ores, Ann. Rep., 1886, Part 4 and Atlas.
Lehigh water gap, Vol. G6.

Lehigh and Wilkes-Barre cross sections, AA Atlas, Northern anthracite, Parts 5 and 6.
Levels above tide in Pennsylvania, Vol. X.
Levels, Butler, Armstrong and Clarion counties, map, Vol. J.
Levels along Slippery Rock creek, map and profile, Vol. J.
Levels, railroads in oil region, Vol. I.2.
Levels, York and Adams counties, Vol. C.
Levels of coal beds in Southwestern Pennsylvania, Ann. Report, 1885.
Lewis colliery cross section, AA Atlas, Southern anthracite, Part 5.
Lias, fossil tracks in, Ann. Rep., 1887.
Ligonier valley coal field, Vols. V2, K3.
Limestones, Susquehanna section, analyses, Vol. M2.
Limestones, Lehigh and Northampton, Vol. D3, Part 1.
Limestones, Cumberland and Lebanon valleys, Ann. Rep, 1886, Part 4 and Atlas.
Limestones, dolomitic, notes on, Vol. M2.
Lincoln colliery columnar section, AA Atlas, Southern anthracite, Part 4b.
Lincoln colliery, cross section, AA Atlas, Southern anthracite, Part 6.
Little Black creek basin, mine maps, etc., AA Atlas, Eastern Middle, Part 1.
Little tunnel, Tamaqua section, AA Atlas, Southern anthracite, Part 4.
Llewellyn colliery mine sheet, AA Atlas, Southern anthracite, Part 2.
Lock Haven, palaeozoic section, Vols. F and G.
Longitude of Wilkes-Barre, Vol. AA.
Lorberry mine sheet, AA Atlas, Southern anthracite, Part 3.
Lower Silurian rocks and sub-divisions summarized, Fin. Sum. Rept. Vol. I.
Lycoming county, Vol. G2.
Lykens mine sheet, AA Atlas, Southern anthracite, Part 3.
Lyon, Shrob & Company's ore lands, Vol. T4.
Luzerne county, Vol. G7.

M.

Map, geological of the State (new), Atlas to Fin. Summary Report.
Map, general, anthracite field, AA Atlas, South anthracite, Part 1.
Map, Southwest Pennsylvania, Ann. Report, 1886, Part 1.
Map, bituminous coal fields of the State, Atlas, Fin. Sum. Report.
Mahanoy basin cross section, AA Atlas, Western Middle, Part 1.
Marshwood colliery columnar section, AA Atlas, Northern anthracite, Part 5.
Marvine colliery columnar section, AA Atlas, Northern anthracite, Part 5.
Marvine colliery cross section, AA Atlas, Northern anthracite, Part 6.
Mauch Chunk formation No. XI, Sum. Fin. Report, Vol. ?, Part 1.
Mayville colliery columnar section, AA Atlas, Western Middle, Part 5.

McKean county, with maps, Vol. RR, Atlas, R2 Atlas.
McKean county oil wells, Vol. I.4.
McCauley mountain, Luzerne county, map and sections, AA Atlas, Eastern Middle, Part 3.
Mercer county, with maps, Vol. Q3.
Mercer county oil wells, Vol. I.4.
Meadville-Sugar Grove section, Vol. I.3, Atlas.
Meadow Brook colliery cross section, AA Atlas, Northern anthracite, Part 6.
Mica, metamorphosis of, Vol. C5, p. 108.
Mineral Spring colliery cross section, AA Atlas, Northen anthracite, Part 6.
Minersville mine sheet, AA Atlas, Southern Anthracite, Part 2.
Middle Creek colliery mine sheet, AA Atlas, Southern anthracite, Part 3.
Middle creek colliery columnar section, AA Atlas, Southern anthracite, Part 4b.
Middle creek colliery cross section, AA Atlas, Southern anthracite, Part 6.
Middleport mine sheet, AA Atlas, Southern anthracite, Part 2.
Mifflin county and map, Vol. F3 and Atlas.
Milford-Coal Hill cross section, AA Atlas, Southern anthracite, Part 5.
Miller's quarry, Hazen coal bank section, Vol. I.3, Atlas.
Minerals, analyses, Vols. B, B2.
Minerals of Delaware county, Vol. C5.
Mines on Monogahela and Youghiogheny rivers, Vol. K4, Ann. Rept., 1885, 1886.
Mining anthracite, methods of, Vol. AC and Atlas.
Mining hematite ores, Vol. D.
Mining methods, Westmoreland Coal Company and Pittsburgh region, Ann. Rep., 1886, Part 1.
Mine Hill colliery section, AA Atlas, Southern anthracite, Part 4.
Monongahela river mines, description, map, plates, Vol. K4.
Montgomery county, southern part, map, Vol. C6.
Montgomery county, central part, map, Vol. C7, Atlas.
Monocraterion, a new fossil, Vol. D2.
Monroe county, Vol. G6.
Montour county, Vol. G7.
Montana, North, sheet, AA Atlas, Western Middle, Part 3.
Moraine terminal, Vol. Z.
Morea colliery, mine sheet, AA Atlas, Southern anthracite, Part 2.
Morea colliery, sections, AA Atlas, Southern anthracite, Part 4.
Morrison's cove, map, Vol. T, Atlas.
Mountain limestone, Ann. Rep., 1885, Vol. H5, preface.
Mountain limestone, Nos. X-XI, Sum. Fin. Rept., Vol. 3, Part 1.
Mt. Carmel, North, sheet, AA Atlas, Western Middle, Part 3.
Mt. Carmel, Map and sections, AA Atlas, Western Middle, Part 2.

Mt. Eagle mine sheet, AA Atlas, Southern anthracite, Part 3.
Mt. Eagle road section, AA Atlas, Southern anthracite, Part 4.
Mt. Jessup colliery columnar section, AA Atlas, Northern anthracite, Part 5.
Mt. Pleasant colliery, mine sheet, AA Atlas, Southern anthracite, Part 2.
Muddy run to Fishing creek section, Lancaster county, Vol. C3, Atlas.
Museum, catalogue of specimens, Vols. O, O2 and O3.

N.

Nanticoke mine sheets and sections, AA Atlas, Northern anthracite, Part 1.
Natural gas, composition and fuel value, Ann. Rep., 1886, Part 2.
Natural gas, use in iron manufacturing, Vol. L.
Neshaminy water basin, maps, Atlas C7.
New Boston mine sheet, AA Atlas, Southern anthracite, Part 2.
New Boston coal basin, Ann. Rep., 1887.
New Castle mine sheet, AA Atlas, Southern anthracite, Part 2.
Newkirk colliery section, AA Atlas, Southern anthracite, Part 4.
New Lincoln colliery section, AA Atlas, Southern anthracite, Part 4.
New Lincoln-Rausch creek cross section, AA Atlas, Southern anthracite, Part 6.
New Philadelphia mine sheet, AA Atlas, Southern anthracite, Part 2.
Newport mine sheets and sections, AA Atlas, Northern anthracite, Part 2.
Newtown mine sheets and sections, AA Atlas, Southern anthracite. Part 2.
New Red formation, geology of, Sum. Fin. Report. Vol. 3, Part 2.
Nittany Valley and Bald Eagle mountain map, Vol. T3.
Notes on Cambria and Somerset counties, H2 and H3, Atlas.
Notes on geological formations in Centre county, Vol. T4.
Northampton county and map, Vol. D3, Part 1, D3, Atlas.
Northern anthracite field, Lackawanna and Wyoming valleys, maps and sections, AA Atlas, Northern anthracite, Parts 1-6.
Northeastern Ohio, comparative geology, Vol. I.
Northdale tunnel cross section. AA Atlas, Southern anthracite, Part 5.
North Diamond colliery cross section, AA Atlas, Southern anthracite. Part 5.
North Montana sheet, map and sections, AA Atlas, Western Middle, Part 3.
North Mt. Carmel sheet, AA Atlas, Western Middle, Part 3.
North Shamokin sheet, AA Atlas, Western Middle, Part 3.
North Shenandoah sheet, map and section, AA Atlas, Western Middle, Part 3.
North vein drift, Ellendale section, AA Atlas, Southern anthracite. Part 6.

Northwest Pennsylvania Summit water basin, map, Vol. I.3, Atlas.
Northwest Pennsylvania, comparative geology, Vol. I.
Northumberland county, Vol. G7.
No. 1 and 2 shaft, cross sections, AA Atlas, Northern anthracite, Part 5.
No. 3 colliery cross section, AA Atlas, Northern anthracite, Part 5.
No. 6 colliery, cross section, AA Atlas, Northern anthracite, Part 5.
No. 18 slope, Wanamie cross section, AA Atlas, Northern anthracite, Part 5.
No. X formation, coal beds in, Ann. Rep., 1885.

O.

Oakdale colliery section, AA Atlas, Southern anthracite, Part 4.
Oakdale-Black mine cross section, AA Atlas, Southern anthracite, Part 5.
Oak Hill colliery cross section, AA Atlas, Southern anthracite, Part 5.
Ohio line, Lawrence county, with maps, Vol. QQ.
Oil district, Vols. I, I.2, I.3, I.4, I.5.
Oil and gas region, general map, Vol. I5.
Oil and gas region, report, Ann. Rep., 1885.
Oil and gas region, report with maps, Ann. Rep., 1886, Part 2.
Oil and gas region, Warren, Venango, Clarion and Butler counties, Vol. I.3.
Oil and gas region, Clarion county, Vol. V2.
Oil production, statistical charts, Vols. I.4, I.5, Ann. Rep., 1886, Part 2.
Oil well records, levels, Vols. I, I.4, I.5.
Oil well records, Butler county chart, Vol. V.
Oil wells completed, etc., statistical chart, Ann. Rep., 1886, Part 2.
Oil wells columnar sections, Vol. I.3, and Atlas.
Oil well rig and tools, Vol. J, Vol. I.3 and Atlas.
Old Carbon Hill colliery cross section, AA Atlas, Northern anthracite, Part 6.
Old Forest City tunnel section, AA Atlas, Northern anthracite, Part 6.
Old river channel, at Packer, map, Vol. V2.
Olyphant mine sheets and sections, AA Atlas, Northern anthracite, Part 4.
Operators, anthracite district, Vol. AA.
Otto-Blackwood cross section, AA Atlas, Southern anthracite, Part 6.
Otto colliery sections, AA Atlas, Southern anthracite, Parts 4-4b.

P.

Paint ores of Lehigh river, Ann. Rep., 1886, Part 4 and Atlas.
Palaeontology of Southwestern Pennsylvania, Vol. K3.
Palaeontology of Perry county, Vol. F2.
Palaeozoic plants, report on, Ann. Rep., 1886, Part 1.
Palmer vein colliery section, AA Atlas, Southern anthracite, Part 4.
Panama conglomerate, Vol. I.3.

Panther creek basin, report. Vol. AA.
Panther creek basin, map, AA Atlas, Southern anthracite, Part 1.
Panther creek mine sheets, AA Atlas, Southern anthracite, Part 1.
Panther creek basin, topographical sheet, AA Atlas, Southern anthracite, Part 1.
Panther creek basin, columnar and cross sections, AA Atlas, Southern anthracite, Part 1.
Parker, map, Armstrong and Butler counties, Vol. V.
Parker, old river channel, map, Vol. V2.
Patterson mine sheet, AA Atlas, Southern anthracite, Part 2.
Payne's colliery section, AA Atlas, Southern anthracite, Part 4.
Peat bog, mineral, at Scranton, Ann. Rep., 1885.
Peckville colliery, columnar section, AA Atlas, Northern anthracite, Part 5.
Pennsylvania Coal Company, 1, 3 and 4 cross section, AA Atlas, Northern anthracite, Part 6.
Pequea-Quarryville section, Lancaster county, C3, Atlas.
Perkiomen water basin, map, C7, Atlas.
Permian flora, Vol. PP.
Perry county and map, Vol. F2.
Persons engaged on survey, 1874-1885, Ann. Rep., 1885, p. xxii.
Petroleum, see oil.
Petroleum, special report on, Vol. J.
Petroleum, genesis of, Vol. I, I.4.
Petroleum, origin of, Vol. K.
Petroleum, general consideration on, Ann. Rep., 1885.
Petroleum, publications on, list of, Ann. Rep., 1886, Part 2.
Philadelphia belt, maps and geology, Vol. C6.
Phoenix Park colliery sections, AA Atlas, Southern anthracite, Part 4, Part 5.
Pierce creek colliery, columnar section, AA Atlas, Northern anthracite, Part 5.
Pike county, Vol. G6.
Pinedale bore hole, columnar section, AA Atlas, Southern anthracite, Part 4b.
Pinedale cross section, AA Atlas, Southern anthracite, Part 5.
Pine Forest colliery section, AA Atlas, Southern anthracite, Part 4.
Pittsburgh coal bed, depths beneath the surface, maps, Vol. K.
Pittsburgh, Boyd's Hill well, record and geological notes, Vol. L.
Pittsburgh coal region, re-survey, Ann. Rep., 1885.
Pittsburgh coal region, report, with maps, Ann. Rep., 1886, Part 1.
Pittsburgh coal region, mining methods, Ann. Rep., 1886, Part 1.
Pittston mine sheets and sections, AA Atlas, Northern anthracite, Part 2.
Plants, palaeozoic, report on, Ann. Rep., 1886, Part 1.
Pleasant valley mine sheets and sections, AA Atlas, Northern anthracite, Part 2.

Plymouth mine sheets and sections, AA Atlas, Northern anthracite. Part 1.
Pocono formation coal beds, Ann. Rep., 1885.
Pocono formation, description, Fin. Sum. Rep., Vol. 3, Part 1.
Pond creek basin, Luzerne county, mine map and sections, AA Atlas, Eastern Middle, Part 2.
Porter and Beach colliery, columnar sections, AA Atlas, Northern anthracite, Part 5.
Potato creek coal basin, map, Vol. R, Atlas.
Pottsville conglomerate, No. XII, Sum. Fin. Report, Vol. 3, Part 1.
Pottsville mine sheets, AA Atlas, Southern anthracite, Part 2.
Pottsville colliery section, AA Atlas, Southern anthracite, Part 4.
Pottsville colliery cross ections, AA Atlas, Southern anthracite, Part 5.
Pottsville gap mine sheet, AA Atlas, Southern anthracite, Part 2.
Pottsville gap section, AA Atlas, Southern anthracite, Part 4.
Pottsville gap, Altamount cross section, AA Atlas, Southern anthracite, Part 5.
Potter county and map, Vol. G3.
Potter county oil wells, Vol. I.4.
Pre-silurian geology, American, Vol. E.
Pre-glacial and post glacial drainage, I.3.
Pre-glacial outlet of Lake Erie, Q4.
Powderly colliery columnar section, AA Atlas, Northern anthracite, Part 5.
Providence mine sheets and sections, AA Atlas, Northern anthracite, Part 3.

Q.

Quaker Hill coal basin, map, Vol. I.4.
Quarternary geology, Wyoming and Lackawanna valleys, Ann. Rep., 1885.
Quinn colliery, cross section, AA Atlas, South anthracite, Part 5.

R.

Radnor township, Delaware county, serpentine beds, Ann. Rep., 1886, Part 4 and Atlas.
Randolph tunnel section, AA Atlas, Southern anthracite, Part 4.
Rattling run mine sheet, AA Atlas, Southern anthraciate, Part 3.
Rattling run tunnel section, AA Atlas, Southern anthracite, Part 4.
Rausch gap, mine sheet, AA Atlas, Southern anthracite, Part 3.
Rausch gap, section, AA Atlas, Southern anthracite, Part 4.
Rausch colliery cross section, AA Atlas, Southern anthracite, Part 6.
Rausch creek-New Lincoln cross section, AA Atlas, Southern anthracite, Part 6.
Records and levals of oil wells, Vol. I.2.

Red Ash colliery cross section, AA Atlas, Southern anthracite, Part 6.
Red Mountain-Kemble cross section, AA Atlas, Southern anthracite, Part 6.
Red beds of the Crawford shale, underground map, Vol. 1.5.
Reevesdale colliery section, AA Atlas, Southern anthracite, Part 4.
Relation of lower carboniferous to Bradford oil sand, Vol. R, Atlas.
Renovo coal basin, Vol. G4.
Repplier colliery cross section, AA Atlas, Southern anthracite, Part 5.
Revenue colliery, cross section, AA Atlas, Southern anthracite, Part 5.
Richardson colliery section, AA Atlas, Southern anthracite, Part 4.
Richardson colliery cross section, AA Atlas, Southern anthracite, Part 5.
Rocktown tunnel, AA Atlas, Southern anthracite, Part 4.
Rogers, H. D., memoir on coal formation, Vol. H, appendix.

S.

Saint Clair mine sheet, AA Atlas, Southern anthracite, Part 2.
Saint Clair shaft section, AA Atlas, Southern anthracite, Part 4.
Salem colliery cross section, AA Atlas, Northern anthracite, Part 5.
Salem colliery cross section, AA Atlas, Southern anthracite, Part 5.
Sandy run basin, Luzerne county, mine map, section, AA Atlas, Eastern Middle, Part 2.
Scranton colliery cross section, AA Atlas, Northern anthracite, Part 6.
Sculptured rocks, Vol. C3.
Seasholtzville ore mine map, Vol. D3, Atlas.
Sections, Butler county, chart, Vol. V.
Sections, columnar, anthracite, Vol. AA.
Sections, Lehigh river, Ann. Rep., 1886. Part 4 and Atlas.
Serpentine beds of Chester and Delaware counties, Vol. C4, p. 346.
Serpentine ranges, Radnor township, Delaware county, Ann. Rep., 1886, Part 4 and Atlas.
Serpentine State line in Chester county, Ann. Rep., 1887.
Shamokin sheet, maps and sections, AA Atlas, Western Middle, Part 2.
Shamokin, North, sheet, AA Atlas, Western Middle, Part 3.
Sharon-Homewood section, map, Vol. V.
Sharp mountain, 11 sections, AA Atlas, Southern anthracite, Part 4.
Shenandoah basin, maps and sections, AA Atlas, Western Middle, Part 1.
Shenandoah basin, North, sheet, AA Atlas, Western Middle, Part 3.
Shenango river, map, Vol. V.
Shenango valley, Vol. V.
Shepherd's creek, mine sheet and sections, AA Atlas, Northern anthracite, Part 4.
Shickshinny mine sheet and sections, AA Atlas, Northern anthracite, Part 2.
Shiro tunnel section, AA Atlas, Southern anthracite, Part 4.
Short mountain colliery section, AA Atlas, Southern anthracite, Part 4.

Short mountain colliery cross section, AA Atlas, Southern anthracite, Part 6.
Sibley colliery cross section, AA Atlas, Northern anthracite, Part 6.
Silver Brook colliery, maps and sections, AA Atlas, Eastern Middle, Part 3.
Silver creek dam, mine sheet, AA Atlas, Southern anthracite, Part 2.
Simpson colliery, columnar section, AA Atlas, Northern anthracite, Part 5.
Slate belt of Northampton and Lehigh, D3, Part 1.
Slippery Rock creek, levels, map and profile, Vol. J.
Slippery Rock creek, map, Vol. V.
Sloane colliery, cross section, AA Atlas, Northern anthracite, Part 6.
Snow Shoe coal basin, Vol. T4.
Snyder county, report and map, Vol. F3 and Atlas.
Somerset county, report, Vol. H3.
Somerset county, map, Atlas to H2 and H3.
Somerset county, Wellersburg basin, Ann. Rep., 1885.
Southwest Pennsylvania map, Ann. Rep., 1886, Part 1.
Southwest Pennsylvania map, oil and gas region, Ann. Rep., 1886, Part 2.
Southern Huntingdon county, cross sections, Vol. F.
South mountain, Durham and Reading Hills, Map 18, sheets, D3 Part 2, D3, Atlas.
South mountain, maps, D5 and D6, Atlas.
South mountain rocks, D3, Part 1.
Somerset county, report and maps, Vol. H3.
Somerset county, oil wells, Vol. I.4.
South Pyne colliery columnar sections, AA Atlas, Southern anthracite, Part 4b.
South Pyne colliery, cross section, AA Atlas, Southern anthracite, Part 6.
South Scranton mine sheets and section, AA Atlas, Northern anthracite, Part 3.
Spangler colliery, cross section, AA Atlas, Southern anthracite, Part 6.
Specimens, catalogue of, in museum, Vols. O, O2 and O3.
Spencer and Milnes colliery, cross section, AA Atlas, Southern anthracite, Part 5.
Spring creek valley, Warren county, topographical map, Vol. I.3, Atlas.
Spring waters, temperatures of, Vol. C3.
Stadia, measurements, theory of, Vol. AA.
State line serpentine, Chester county, Ann. Rep., 1887.
Stone mountain fault, Vol. T3.
Sub-carboniferous section, Vol. G4.
Sub-conglomerate, Bradford oil district, Vol. R, Atlas.
Sugar Grove-Meadesville section, Vol. I.3, Atlas.
Sugar-Notch mine sheets and sections, AA Atlas, Northern anthracite, Part 1.

Sugar Notch and Maffett cross sections, AA Atlas, Northern anthracite, Part 5.
Sullivan county, Vol. G2.
Susquehanna river section through Wrightsville, Vol. C.
Susquehanna river section, map, six sheets, Vol. C3, Atlas.
Susquehanna river region, six counties, G7.
Susquehanna section, sub-carboniferous, Vol. G4.
Susquehanna county, Vol. G5.
Swatara mine sheet, AA Atlas, Southern anthracite, Part 2.

T.

Tamaqua basin, mine sheet, AA Atlas, Southern anthracite, Part 2.
Tamaqua basin, map, AA Atlas, Southern anthracite, Part 5.
Tamaqua basin, sections, AA Atlas, Southern anthracite. Part 4.
Tamaqua basin, cross section, AA Aalas, Southern anthracite, Part 5.
Tangascootack coal basin, Vol. G4.
Temperature of spring water in Lancaster county, Vol. C3.
Terminal moraine, Vol. Z.
Terraces, Monongahela river, Vol. Z, p. x.
Third sand, Venango, map, Vol. I.3, Atlas.
Thomaston colliery, section, AA Atlas, Southern anthracite, Part 4.
Thomaston colliery, cross section, AA Atlas, Southern anthracite, art 5.
Tioga county and map, Vol. G.
Tipton run coal, Blair county, Ann. eRp., 1885.
Tohickon water basin, map, Atlas C7.
Tomhickon sheet, Humboldt, Mt. Pleasant, etc., maps, sections, AA Atlas, Eastern Middle, Part 3.
Tonnage, anthracite, 1874-1882, Vol. AA.
Topographical survey recommended, Vol. H6, p. vii.
Topographical survey, Ann. Rep., 1885, p. xxxiii.
Topographical sheets, Southern anthracite field, AA Atlas, Southern anthracite, Part 1.
Topographical sheets, Reading-Durham hills, D3. Atlas.
Tower City, mine sheet, AA Atlas, Southern anthracite. Part 3.
Tracks, fossil in Trias, Ann Rep., 1887.
Transit line to establish altitudes, notes, Vol. C.
Tremont mine sheet, AA Atlas, Southern anthracite, Part 3.
Trevorton sheet, maps and section, AA Atlas, Western Middle, Part 2.
Triangulation, geodetic report, Ann. Rep., 1885.
Tuscarora mine sheet, AA Atlas, Southern anthracite, Part 2.

U.

Union county, report and map, Vol. F3 and Atlas.
Upper Barren Measures, columnar section, Vol. I.3, Atlas.
Upper Silurian and Devonian measures of the State described, Fin. Sum, Rep., Vol. II.
Upper Lehigh basin, Luzerne county, mine maps, sections, AA Atlas, Eastern Middle, Part 2.

V.

Vegetable origin of coal, Ann. Rep., 1885.
Venango county map, Vol. I.3, Atlas.
Venango county, oil district with maps, Vols. I, I.3.
Venango Third sand, map and section, Vol. I.3, Atlas.
Venango-Butler oil group vertical section, I.5.
Voorhees colliery cross section, AA Atlas, Southern anthracite, Part 5.

W.

Wadesville shaft section, AA Atlas, Southern anthracite, Part 4.
Wadesville colliery cross section, AA Atlas, Southern anthracite, Part 5.
Walker bore holes, columnar sections, AA Atlas, Northern anthracite, Part 5.
Wanamie cross sections, AA Atlas, Northern anthracite, Part 5.
Warrantee maps, Elk, Forest, parts of Clarion and Jefferson, R2, Atlas.
Warren county with map, Vol. I.4.
Warren county, oil region and wells, Vol. I.3, I.4.
Warren, geology of vicinity, Vol. I.
Warrior run, mine sheets and sections, AA Atlas, Northern anthracite, Part 1.
Warrior run, cross section, AA Atlas, Northern anthracite, Part 5.
Washington county, report and map, Vol. K.
Washington and Greene bituminous district and maps, Vol. K, Ann. Rep., 1886.
Washing hematite ores, Vol. D.
Water basins summit in Northwestern Pennsylvania, map, Vol. I.3, Atlas.
Water course, ancient, near Franklin, Pa., map, Vol. I.3, Atlas.
Water Department of Philadelphia, survey and maps, Vol. C7, Atlas.
Water gaps, D3, Part 1.
Wayne county, report and map, Vol. G5.
Waynesburg coal bed, depths beneath surface, maps, Vol. K.

Weatherly, sheet, Carbon county, AA Atlas, Eastern Middle, Part 2.
Wellersburg basin, coals and fire clays, Ann. Rep., 1885.
Wellsburg section revised, H2, H3, Atlas.
Welsh company tunnel sections, AA Atlas, Southern anthracite, Part 4.
West Virginia carboniferous formation, Vol. P2.
West West basin, mine sheet, AA Atlas, Southern anthracite, Part 2.
Westwood colliery, mine sheet, AA Atlas, Southern anthracite, Part 2.
Westwood colliery columnar sections, AA Atlas, Southern anthracite, Part 4b.
West Buck mountain colliery, maps, sections, AA Atlas, Eastern Middle, Part 3.
West Delaware colliery cross section, AA Atlas, Southern anthracite, Part 5.
Western Middle coal field, Mahanoy and Shenandoah basins, maps and sections, AA Atlas, Western Middle, Parts 1, 2, 3.
Western New York, comparative geology, Vol. I.
Westmoreland county, west of Chestnut ridge, map, Vol. K2.
Westmoreland county, Ligonier valley, map, Vol. K3.
Westmoreland county, mining methods, Ann. Rep., 1886, Part 1.
White Bridge colliery, columnar sections, AA Atlas, Northern anthracite, Part 5.
Whitfield's slope cross section, AA Atlas, Southern anthracite, Part 5.
Wilkes-Barre, latitude and longitude, Vol. AA.
Wilkes-Barre mine sheets and sections, AA Atlas, Northern anthracite, Part 1.
Williamstown colliery mine sheet, AA Atlas, Southern anthracite, Part 3.
Williamstown colliery, sections, AA Atlas, Southern anthracite, Parts 4, 4b, 6.
Windy Harbor-Bear Ridge cross section, AA Atlas, Southern anthracite, Part 5.
Winton colliery, columnar section, AA Atlas, Southern anthracite, Part 5.
Wisconisco, West End, mine sheet, AA Atlas, Southern anthracite, Part 3.
Wolf creek colliery section. AA Atlas, Southern anthracite, Part 4.
Woods colliery section. AA Atlas, Southern anthracite, Part 4.
Wrightsville section along the Susquehanna, Vol. C.
Wyoming county. Vol. G7.
Wyoming basin, western end mine sheets 1 and 2, Ann. Rep., 1886. Part 3, Atlas.
Wyoming colliery cross section, AA Atlas, Northern anthracite, Part 5.
Wyoming valley, Quarternary geology, Ann. Rep., 1885.
Wyoming valley, limestone beds and fossils, Ann. Rep., 1885.
Wyoming buried valleys, Ann. Rep., 1885.

Y.

Yatesville colliery mine sheets and sections, AA Atlas, Northern anthracite, **Part 2.**

Yatesville colliery, cross section, AA Atlas, Northern anthracite, Part 6.

Yellow Springs colliery mine sheet, AA Atlas, Southern anthracite, **Part 3.**

Yellow Springs tunnel section, AA Atlas, Southern anthracite, Part 6.

York county, report and map, Vols. C, C2, C3 and Atlas.

York Farm-East Pine Knot cross section, AA Atlas, Southern anthracite, **Part 5.**

Yorkville colliery mine sheet, AA Atlas, Southern anthracite, Part 2.

Youghiogheny river mines, description, with map, Vols. K4, L.

Youghiogheny valley, iron ores, fire clays, cement, sand and salt, Vol. L.